Discrete Mathematics for Computer Science

Discrete Mathematics for Computer Science

An Example-Based Introduction

Jon Pierre Fortney

CRC Press
Taylor & Francis Group
Boca Raton London New York

CRC Press is an imprint of the
Taylor & Francis Group, an **informa** business

A CHAPMAN & HALL BOOK

First edition published 2021
by CRC Press
6000 Broken Sound Parkway NW, Suite 300, Boca Raton, FL 33487-2742

and by CRC Press
2 Park Square, Milton Park, Abingdon, Oxon, OX14 4RN

Library of Congress Cataloging-in-Publication Data

Names: Fortney, Jon Pierre, author.
Title: Discrete mathematics for computer science : an example-based introduction / Jon Pierre Fortney.
Description: First edition. | Boca Raton : C&H/CRC Press, 2021. | Includes bibliographical references and index.
Identifiers: LCCN 2020034984 (print) | LCCN 2020034985 (ebook) | ISBN 9780367549886 (hardback) | ISBN 9781003091479 (ebook).
Subjects: LCSH: Computer science--Mathematics.
Classification: LCC QA76.9.M35 F67 2021 (print) | LCC QA76.9.M35 (ebook) | DDC 004.01/51--dc23
LC record available at https://lccn.loc.gov/2020034984
LC ebook record available at https://lccn.loc.gov/2020034985

ISBN: 978-0-367-54988-6 (hbk)
ISBN: 978-0-367-54989-3 (pbk)
ISBN: 978-1-003-09147-9 (ebk)

Typeset in Computer Modern font
by KnowledgeWorks Global Ltd.

To Ron Ryan Noval

Contents

Preface

One of the major challenges of teaching mathematics at the community college or small liberal arts college level is finding books that are genuinely appropriate for the student population. Writing this book for the discrete mathematics for computer science course at Zayed University happened because I was completely unable to find any textbook on the market that both (1) covered the material I needed to cover and (2) was written at a level appropriate for my students. Often textbooks cover a huge range of material, usually far more material than even good students can fully assimilate and understand in a single semester. On some level this is not surprising; textbooks are written by experts in the topic who deeply understand the nuances of the field and want to make everything as precise and "clear" as possible, while still illustrating the subtleties and complexities of the topic. Unfortunately, the meaning of the word "clear" depends on the individual. Books that are extremely precise and that delve into subtleties or nuance are completely inappropriate for many college level students. I believe students benefit most with a clear, clean, uncluttered, down-to-earth exposition of the material.

This book is intended for a first- or second-year discrete mathematics course for computer science majors. In it I cover many of the most important mathematical topics that future computer science majors need to be aware of. The first chapter introduces algorithms. As an understanding of algorithm basics is essential for computer science majors, I have integrated simple algorithms throughout the book. Chapter two covers number representations and converting between decimal, binary, octal, and hexadecimal numbers. Chapter three on logic covers propositions, connectives, truth tables, and the laws of logic. Chapter four covers all the basic definitions of set theory, Venn diagrams, and the laws of set theory. Chapter five on Boolean algebra introduces the laws of Boolean Algebra and the application of Boolean algebra to digital circuits. This chapter also ties together logic, set theory, and Boolean algebra. Chapter six introduces functions, and in particular, introduces the set-theoretic definition of a function. Chapter seven on counting and combinatorics covers the addition and multiplication principles, permutations and combinations. Chapter eight uses what was learned in chapters one, six, and seven to introduce algorithmic complexity and big-O notation. Chapter nine covers basic graph theory, Euler and semi-Euler graphs, and matrix representations for graphs. Chapter ten introduces trees and then covers minimal spanning trees and minimal distance paths. A variety of the covered topics are utilized in appendix A to design a circuit to add two eight-digit binary numbers. Appendix B contains the answers to all the end-of-chapter problems in the book.

This book has several features that make it essentially unique among discrete mathematics books, particularly discrete mathematics books aimed at computer science majors. It is written with students at community colleges or small liberal arts colleges in mind. The language in this book is very straightforward. Sentences are not complex and, as much as possible, the vocabulary and explanations are kept simple. Whenever definitions or explanations are made, they are quickly followed by examples to aid students in their understanding. Words being defined are always given in bold face. Paragraphs are kept short and the amount of prose a student is required to read between examples is kept to a minimum. All of this not only strongly reinforces the definition or explanation, but also aids students to fully understand the written content. Indeed, this textbook would be ideal for international students and non-native English speakers. Furthermore, students without much mathematical maturity or abstract thinking ability need numerous examples to help them bridge the gap between concrete and abstract thought. They also need a significant amount of practice to help them both learn and internalize the material. I have attempted to provide both numerous examples and practice problems for students to do just this. There are over 200 worked examples scattered throughout the book, boxed for easy reference. There are also over 200 end-of-chapter problems, many of which include multiple parts, which provide students ample practice opportunities. The answers to these problems are provided in appendix B, allowing students to check their work.

Another strategy I have used is to relate the mathematical topics back to computer science as much as is reasonably possible. At the level the textbook is written, many of the examples drawn from computer science are necessarily quite superficial and simple, but they still provide students interested in computer science motivation to understand the mathematics. Making the mathematics relevant to their future classes and lives is essential in encouraging students to learn the material. Explaining the importance and relevance of algorithms in Chapter one and then integrating algorithms throughout the rest of the material both gives students the opportunity to practice and better understand algorithms, and gives them a computer science-based approach to interacting with and learning other mathematical topics. There are about 40 algorithms integrated into the text, highlighted with a light gray background for easy reference. These algorithms form an integral part of the text and help the students learn how to read and understand pseudocode. While logic, set theory, and Boolean algebra all have their relevance to computer science, they are also all deeply related. In the chapter on Boolean algebra the relationship between these topics is explored. Similarly, the connection between functions, algorithms, and counting principles is considered in the chapter on algorithmic complexity. The chapters on graph theory and trees have numerous applications in computer science, some of which are considered, along with the algorithms that implement them. And finally, appendix A explores the relation between binary numbers, Boolean algebra, and circuit design.

As written, I believe the entire book can be easily covered in a single semester. If used in a school that follows the quarter system, if necessary, Chapters seven and ten could be omitted without compromising the mathematical material too much. For example, when teaching in the summer I generally only cover the first two sections of Chapter seven and the first section of Chapter ten.

Finally, I would like to express my appreciation to both Ron Noval and Rene Hinojosa for their ongoing support throughout the writing of this book. I would also like to thank my many past students who offered invaluable comments on the various drafts of this book and who pointed out innumerable typos.

Jon Pierre Fortney
June 2020

CHAPTER 1

Introduction to Algorithms

Algorithms play an extremely important role in both computer science and mathematics. Algorithms are detailed instructions on how to carry out some specific task. Computer programs are implementations of algorithms, so it is very important that computer science majors and programmers have a good understanding of how algorithms work. Deciding on an algorithm is generally the first step of writing a computer program. Because algorithms will be used throughout this book we begin by introducing some of the basic concepts and ideas related to algorithms.

1.1 WHAT ARE ALGORITHMS?

An **algorithm** is a finite sequence of **well-defined** steps to perform some task such that

- each step is a clear instruction that can be done in a finite amount of time,
- the sequence in which the steps are done is clear, and
- the process is guaranteed to stop after a finite number of steps have been done.

Here the phrase *well-defined* means the steps are very clear. There is no confusion or uncertainty in what is meant. A computer program is the **implementation** of an algorithm in some programming language. This means the program actually carries out, or **executes**, the algorithm.

Designing an algorithm is one of the first steps in writing a computer program. In order to make algorithms easier for us to understand they are written in a form of English called **pseudocode**. This allows us to concentrate and think about the structure of the algorithm without worrying about the details of a particular computer language. Let us start by looking at a very simple algorithm written in pseudocode. Here is an algorithm to calculate the volume of a cube.

Algorithm: Finds the volume of a cube with side length l.

1. Input l. (l is the length of one side of the cube.)
2. *volume* $\longleftarrow l^3$
3. Output *volume*.

Notice the following things about this algorithm:

- The steps are numbered for easy reference.
- Comments explaining something are put in parenthesis.
- The symbol $\longleftarrow$ means **assignment**. Here the formula l^3 is evaluated and the resulting number is **assigned** to the variable called *volume*.

In algebra we may write $x = 7$. The x is a variable name and by saying $x = 7$ we are giving the variable the value 7. In algebra variables are usually just letters, like x, y, or z. But in computer science variables are often given names. This helps programmers remember what the variable means. So, *volume* $\longleftarrow$ l^3 means that the variable named *volume* is given whatever value l^3 is. In other words, the number given by l^3 is stored in, or assigned to, the computer's memory at an address that is called *volume*.[1]

1.2   CONTROL STRUCTURES

Algorithms use something called **control structures** to tell the algorithm what to do at different times. These are instructions that control how the algorithm works. For example, they tell how many times other instructions are executed, or when they are executed. There are two kinds of control structures, **conditional controls** and **loop controls**. Conditional controls give conditions for the execution of an algorithm step. Conditional controls include

- **if-then**,
- **if-then-else**.

Loop controls tell how many times a step in an algorithm should be executed. Loop controls include

- **for-do**,
- **while-do**,
- **repeat-until**.

We look at each of these control structures in turn.

If-then

Consider the following simple algorithm that contains an **if-then** conditional control. Notice that each line, or step, is numbered.

Algorithm: Determines if a student has passed.

1. Input x. (x is a student's grade as a percent.)
2. If $x \geq 60$ **then**
   2.1 Output "passed."

In line 1 the algorithm asks you to input a student's grade as a percentage. Line 2 says that **if** the condition is true and the grade is greater than or equal to 60 **then** move to line 2.1, which outputs the word *passed*. Suppose you input a 75. Since the grade is $75 \geq 60$ then the word *passed* is output and the algorithm ends. But suppose the grade was 55. Then since $55 \not\geq 60$ we do not go to line 2.1 and the algorithm ends with nothing output.

---

[1] In computer science when one sees an equal sign then that usually means that one is comparing two objects to see if they are the same. This is addressed more in Chapter 8. An equal sign could also mean that one is defining two object to be the same.

If-then-else

Now consider the following algorithm. It contains an **if-then-else** conditional control.

Algorithm: Determines if a student has passed or failed.

1. Input x. (x is a student's grade as a percent.)
2. **If** $x \geq 60$ **then**
   2.1 Output "passed."
   else
   2.2 Output "failed."

In line 1 the algorithm asks you to input a student's grade as a percentage. Line 2 says that **if** the grade is greater than or equal to 60 **then** move to the next line, 2.1, which outputs the word *passed*. But now there is an **else**. This tells the algorithm what to do if the condition is not met. Suppose you input a 75. Then since $75 \geq 60$ the algorithm goes to line 2.1, outputs the word *passed*, and ends. But now suppose that a grade of 55 was input. Since $55 \ngeq 60$ then the algorithm goes to line 2.2, outputs the word *failed*, and ends. Since the **if** condition was not met the **then** line was skipped and the algorithm goes to the **else** line.

It is also possible to **nest** **if-then-else** statements. This means we put an **if-then-else** statement inside another **if-then-else** statement. Consider the following algorithm that prints out a student's letter grade when their grade as a percent is input.

Algorithm: Gives a student's letter grade.

1. Input x. (x is a student's grade as a percent.)
2. **If** $x \geq 90$ **then**
   2.1 Output "A."
   else
   3. **If** $x \geq 80$ **then**
      3.1 Output "B."
      else
      4. **If** $x \geq 70$ **then**
         4.1 Output "C."
         else
         5. **If** $x \geq 60$ **then**
            5.1 Output "D."
            else
            5.2 Output "F."

Here we can see several nested **if-then-else** statements. Consider what happens if we input the grade 75. In line 2 the first **if** condition is checked and $75 \ngeq 90$ is found so line 2.1 is skipped and we go to line 3. In line 3 the second **if** condition is checked and $75 \ngeq 80$ is found so line 3.1 is skipped and we go to line 4. In line 4 the third **if** condition is checked and $75 \geq 70$ is found so we go to line 4.1 and output the letter grade C. At this point the algorithm ends.

For-do

Now we look at loop controls. Loop controls tell how many times a step in the algorithm should be executed. We can think of the algorithm as looping around and repeating the same step a number of times. First we look at the **for-do** loop control.

Algorithm: Finds the smallest number in a list of numbers.

1. Input the number of values in list, n.
2. Input the list of numbers $x_1, x_2, \ldots, x_n$.
3. $min \longleftarrow x_1$
4. $i \longleftarrow 2$
5. **For** $i = 2$ to n **do**
   5.1 **If** $x_i < min$ **then**
       5.1.1 $min \longleftarrow x_i$
   5.2 $i \longleftarrow i + 1$
6. Output min.

The number of values in the list is input in step 1, then the list of values is input in step 2. In step 3 the word min is the name of a variable and x_1, the first value in the list, is assigned to it. Step 4 **initializes** the variable i with 2. This means the variable i is given the starting, or initial, value 2. Step 5 contains the **for-do** loop. This line tells us we repeat steps 5.1, 5.1.1 if necessary, and 5.2 for each value of i between 2 and n. Step 5.2 increases i by 1. This is often called **incrementing** i. Once i becomes $n + 1$ we no longer repeat step 5 and we move onto step 6 where we output the minimum value.

Remember, pseudocode is written in English at a level that allows us to fully understand the algorithm. As long as we can fully understand what is supposed to happen it is enough. Suppose we simplify this algorithm a little by taking out steps 4 and 5.2, the steps that involve initializing and incrementing the variable i. Then the algorithm looks like this.

Algorithm: Finds the smallest number in a list of numbers.

1. Input the number of values in list, n.
2. Input the list of numbers $x_1, x_2, \ldots, x_n$.
3. $min \longleftarrow x_1$
4. **For** $i = 2$ to n **do**
   4.1 **If** $x_i < min$ **then**
       4.1.1 $min \longleftarrow x_i$
5. Output min.

In **for-do** loops we step through values of i. The $i = 2$ to n inside the **for-do** loop tells us for which values of i we need to do the loop. We start with $i = 2$ and each time the loop is executed the value of i increases by one. We keep executing the loop until we reach the last value of i, which is n. When the i value is incremented to $n + 1$ we stop doing the loop. Hopefully this should be clear to any human reading this algorithm and so we do not really need those other steps to understand what is happening.

While-do

The **while-do** loops are a lot like the **for-do** loops. **While-do** loops are repeated while a given condition is true. In other words, as long as the given condition is true the loop repeats. Consider the following algorithm.

Algorithm: Finds the order of the factor two for the integer n.

1. Input a non-negative integer n.
2. $order \longleftarrow 0$
3. **While** n is even **do**
    3.1. $n \longleftarrow \frac{n}{2}$
    3.2. $order \longleftarrow order + 1$
4. Output $order$.

In this example the steps 3.1 and 3.2 are repeated as long as n is an even number. At some point the condition stops being true and the loop is no longer repeated.

Repeat-until

The **repeat-until** loops are very similar to the **while-do** loops. **While-do** loops are repeated **while** a given condition is true. **Repeat-until** loops are repeated **until** a given condition is true. That means the loop repeats while the condition is false and then stops when the condition becomes true. In this example the steps 2.1 and 2.2 are repeated until the condition $i = 0$ is met.

Algorithm: Outputs a list of the numbers from n to 1.

1. Input a positive integer n
2. $i \longleftarrow n$.
3. **Repeat**
    3.1. Output i
    3.2. $i \longleftarrow i - 1$
    Until $i = 0$

1.3   TRACING AN ALGORITHM

One of the techniques often used to help us understand how an algorithm works is to **trace** it. We trace an algorithm by choosing some particular input and then step through the algorithm and write down the values of all the variables at each step until the algorithm ends. In other words, the trace of an algorithm tell us what value every variable in the algorithm has at each step. By tracing an algorithm for different possible inputs we understand what an algorithm does and how it works much better. To learn how tracing an algorithm works we will first trace the three algorithms given as examples of loop controls.

Algorithm: Finds the smallest number in a list of numbers.

1. Input the number of values in list, n.
2. Input the list of numbers $x_1, x_2, \ldots, x_n$.
3. $min \longleftarrow x_1$
4. **For** $i = 2$ to n **do**
    4.1 **If** $x_i < min$ **then**
        4.1.1 $min \longleftarrow x_i$
5. Output min.

Let us suppose we want to find the smallest number in the list $5, 4, 8, 3$. There are four numbers in this list so $n = 4$ and $x_1 = 5, x_2 = 4, x_3 = 8, x_4 = 3$. We usually use a table to trace an algorithm.

Step	min	i	x_i	output
3	5	-	-	-
4	5	2	4	-
4.1	5	2	4	-
4.1.1	4	2	4	-
4	4	3	8	-
4.1	4	3	8	-
4	4	4	3	-
4.1	4	4	3	-
4.1.1	3	4	3	-
5	3	4	3	3

Often when we trace an algorithm we do not bother to write down the input steps in the table. Here we did not write steps 1 or 2 but started with step 3 where the value $x_1 = 5$ was assigned to the variable min. After step 3 we included a row in the table for each of the following steps in the algorithm. In step 4 the algorithm states "**For** $i = 2$ to n **do**" so we let $i = 2$. If $i = 2$ then we also know what x_i is and so also include that in the table. Step 4.1 states "**If** $x_i < min$ **then**" so we check to see if $4 < 5$. Since it is, we go to the **then** part of the **if-then** statement, step 4.1.1. Here $x_2 = 4$ is assigned to the variable min. After that we return to the next step in the **for-do** statement where $i = 3$. Notice that this time around x_3 is not less than min and so step 4.1.1 is skipped. The algorithm continues the **for-do** loop until $i = n$. Carefully follow the rest of the steps of the algorithm to make sure you understand them.

Now we will look at the algorithm that contains the **while-do** loop.

Algorithm: Finds the order of the factor two for the integer n.

 1. Input a non-negative integer n.
 2. $order \longleftarrow 0$
 3. **While** n is even **do**
      3.1. $n \longleftarrow \frac{n}{2}$
      3.2. $order \longleftarrow order + 1$
 4. Output $order$.

We will trace this algorithm for the non-negative integer $n = 56$.

Step	$order$	n	output
2	0	56	-
3	0	56	-
3.1	0	28	-
3.2	1	28	-
3	1	28	-
3.1	1	14	-
3.2	2	14	-
3	2	14	-
3.1	2	7	-
3.2	3	7	-
3	3	7	-
4	3	7	3

As before we skip the input step and begin the trace of the algorithm with step 2. In step 2 we assign the value 0 to *order*. In step 3 we check to see if n is even. Since it is we do steps 3.1 and 3.2. In step 3.1 the number n is divided by two and this new number is assigned to n. Then step 3.2 adds one to *order* and assigns this new number to *order*. We could also say we increase the value of *order* by one. Then we return to step 3 where we check if the new value of n is even. **While** n is still even we **do** steps 3.1 and 3.2. When n stops being even we skip steps 3.1 and 3.2 and continue with the algorithm.

Now look at the two algorithm traces we presented. Notice that in both cases a number of lines look exactly the same. In the first trace all the lines 4 and 4.1 are exactly the same. In the second trace when 3 follows a 3.2 then these two lines are the same. Often when one gets good at tracing algorithms one skips writing down repeated lines.

Algorithm: Outputs a list of numbers going from n to 1.

1. Input a positive integer n.
2. $i \leftarrow n$.
3. **Repeat**
    3.1. Output i
    3.2. $i \leftarrow i - 1$
    Until $i = 0$

This is a simple algorithm that shows how the **repeat-until** loop control works. We will trace it for $n = 3$.

Step	i	output
2	3	-
3.1	3	3
3.2	2	-
3.1	2	2
3.2	1	-
3.1	1	1
3.2	0	-

Tracing this algorithm is straightforward. As long as the condition after **until** is not met we continue to **repeat** steps 3.1 and 3.2. In step 3.1 we output i and in step 3.2 we subtract one from i and assign the new value to i. This is called **decrementing** i and works very similarly to incrementing i, only instead of adding one each time we subtract one each time.

Learning how to trace an algorithm is a very important skill. Tracing an algorithm multiple times for different inputs is a very good way to help you understand what an algorithm does and how it works. As you get better at tracing algorithms you will probably find yourself skipping steps or only writing down enough detail to understand what is happening. This is fine. Tracing algorithms is simply a tool to help you understand an algorithm. Your understanding of how an algorithm works is the most important thing.

1.4 ALGORITHM EXAMPLES

Now we will simply look at a number of different algorithms to get some practice in tracing and trying to understand algorithms. Remember, tracing an algorithm is a tool to help you understand how an algorithm works and what it does. Understanding the algorithm and how it behaves is the most important part. We begin with an algorithm that is an example of the **if-then-else** conditional control.

Algorithm: Finds the larger of two numbers.

1. Input x and y.
2. **If** $x > y$ **then**
   2.1 $max \longleftarrow x$
  else
   2.2 $max \longleftarrow y$
3. Output max.

Example 1.1

Use the above algorithm to find the larger of the two numbers $x = 7$ and $y = 9$.

This algorithm is so simple that tracing it in a table is a little silly. Simply step through each line of the algorithm. In line 1 the values $x = 7$ and $y = 9$ are input in line 1. In line 2 we check if $x > y$. Since $7 \not> 9$ then we skip line 2.1 and do line 2.2 where we assign 9 to max. Then in line 3 we output max which is 9.

Next we look at an algorithm that is an example of the **for-do** loop control.

Algorithm: Evaluates x^n where x is a real number and n is a positive integer.

1. Input x and n.
2. $answer \longleftarrow x$
3. **For** $i = 1$ to $n - 1$ **do**
   3.1. $answer \longleftarrow answer \times x$
4. Output $answer$.

Example 1.2

Trace the above algorithm for $x = 7$ and $n = 4$. In other words, use the algorithm to find 7^4.

Step	i	$answer$	Output
2	-	7	-
3.1	1	49 ($\longleftarrow 7 \times 7$)	-
3.1	2	343 ($\longleftarrow 49 \times 7$)	-
3.1	3	2401 ($\longleftarrow 343 \times 7$)	-
4	3	2401	2401

Here we look at an algorithm is also an example of the **while-do** loop control. The exclamation mark ! following a positive integer is called the **factorial** symbol. It will be covered in chapter 7. Simply put, $n!$ means to multiply n by all the numbers less than n. For example,

$$2! = 2 \times 1,$$
$$3! = 3 \times 2 \times 1,$$
$$4! = 4 \times 3 \times 2 \times 1,$$

and so on.

Algorithm: Evaluates $n!$ where n is a positive integer.

1. Input positive integer n.
2. *answer* $\longleftarrow n$
3. **While** $n > 1$ **do**
   3.1 $n \longleftarrow n - 1$
   3.2 *answer* $\longleftarrow$ *answer* $\times n$
4. Output n.

Example 1.3

Trace the above algorithm for $n = 4$. In other words, use the algorithm to find $4!$.

Step	n	*answer*	Output
2	4	4	-
3.1	3	4	-
3.2	3	12 ($\longleftarrow 4 \times 3$)	-
3.1	2	12	-
3.2	2	24 ($\longleftarrow 12 \times 2$)	-
3.1	1	24	-
3.2	1	24 ($\longleftarrow 24 \times 1$)	-
4	1	24	24

The next few algorithms are examples of search algorithms.

Algorithm: Searches a string of integers $x_1, x_2, \ldots, x_n$ to see if it contains the integer s.

1. Input string $x_1, x_2, \ldots, x_n$ and integers n and s.
2. $i \longleftarrow 1$
3. *integer_s_detected* $\longleftarrow$ *false*
4. **Repeat**
   4.1. **If** $x_i = s$ **then**
      4.1.1. *integer_s_detected* $\longleftarrow$ *true*
   4.2. $i \longleftarrow i + 1$
   until *integer_s_detected* $=$ *true* or $i = n + 1$
5. **If** *integer_s_detected* $=$ *true* **then**
   5.1 Output "String contains s."
   else
   5.2 Output "String does not contain s."

Here the steps 4.1, 4.1.1, and 4.2 are repeated until either *integer_s_detected* $=$ *true* or until $i = n + 1$. Thus, as soon as we have $x_i = s$ in step 4.1 then step 4.1.1 is called and *integer_s_detected* becomes true. Step 4.2 is still carried out but then the two conditions are checked it see if we need to repeat step 4 again or not. We do not need to continue checking the rest of the string once we have found $x_i = s$ for some value of i. Finally, an **if-then-else** statement is used in order to output the correct statement.

Example 1.4

Use the above algorithm to search the string 758 to see if it contains a 5.

The input is $x_1 = 7$, $x_2 = 5$, $x_3 = 8$, $n = 3$, and $s = 5$.

Step	*integer_s_detected*	i	x_i	Output
2	-	1	-	-
3	*false*	1	-	-
4.1	*false*	1	7	-
4.2	*false*	2 ($\leftarrow 1 + 1$)	7	-
4.1	*false*	2	5	-
4.1.1	*true* ($\leftarrow$ *true*)	2	5	-
4.2	*true*	3 ($\leftarrow 2 + 1$)	5	-
5.1	*true*	3	5	String contains 5.

This algorithm works fine, but we can make it much simpler by using a **return**. Return works a lot like Output except that it also ends the algorithm. Sometimes we want an algorithm to end as soon as something is output. That is when return is used.[2] In this example, as soon as we encounter an integer s in the string we know the string contains the integer s. We do not need to continue checking the rest of the string. Thus, if $x_i = s$ for some value of i then "String contains s" is output and the algorithm ends right there.

> **Algorithm:** Searches a string of integers $x_1, x_2, \ldots, x_n$ to see if it contains the integer s.
>
> 1. Input string $x_1, \ldots, x_n$, n and s.
> 2. **For** $i = 1$ to n **do**
>   2.1. **If** $x_i = s$ **then**
>   2.1.1. Return "String contains s."
> 3. Return "String does not contain s."

Example 1.5

Trace the above algorithm for the string 497316 and search for the integer 3.

The input is $x_1 = 4$, $x_2 = 9$, $x_3 = 7$, $x_4 = 3$, $x_5 = 1$, $x_6 = 6$, $n = 6$, and $s = 3$.

Step	i	x_i	s	Output
2	1	1	-	-
2.1	1	4	3	-
2.1	2	9	3	-
2.1	3	7	3	-
2.1	4	3	3	-
2.1.1	4	3	3	String contains 3.

Notice, since $x_4 = s$ the **then** part of the **if-then** statement is executed, "String contains 3." is output, and the algorithm ends.

---

[2]This is actually a simplification, the return statement exits the function in which it occurs, either with or without a value, and returns control to the calling function.

Algorithm: Searches two strings of integers $x_1, x_2, \ldots, x_n$ and $y_1, y_2, \ldots, y_n$ of equal length to see if either string contains integer s.

1. Input strings $x_1, \ldots, x_n$ and $y_1, \ldots, y_n$, n and s.
2. **For** $i = 1$ to n **do**
    2.1. **If** $x_i = s$ **then**
        2.1.1. Return "String one contains s."
3. **For** $i = 1$ to n **do**
    3.1. **If** $y_i = s$ **then**
        3.1.1. Return "String two contains s."
4. Return "Neither string contains s."

Example 1.6

Use the above algorithm to see if either string 74 or string 89 contains an 8.

The input is $x_1 = 7$, $x_2 = 4$, $y_1 = 8$, $y_2 = 9$, $n = 2$, and $s = 8$.

Step	i	x_i or y_i	Output
2.1	1	7	-
2.1	2	4	-
3.1	1	8	-
3.1.1	1	8	String two contains 8.

Algorithm: Checks if two strings $x_1, \ldots, x_n$ and $y_1, \ldots, y_n$ contain a common integer. (That is, checks to see if there is an integer that is contained in both strings.)

1. Input strings $x_1, \ldots, x_n$ and $y_1, \ldots, y_n$, and length n.
2. **For** $i = 1$ to n **do**
    2.1 **For** $j = 1$ to n **do**
        2.1.1 **If** $x_i = y_j$ **then**
            2.1.1.1 Return "Element in common to both strings."
3. Return "No elements in common to both strings."

Example 1.7

Use the above algorithm to see if the strings 67 and 89 have any integers in common.

The input is $x_1 = 6$, $x_2 = 7$, $y_1 = 8$, $y_2 = 9$, and $n = 2$.

Step	i	j	x_i	y_j	Output
2.1.1	1	1	6	8	-
2.1.1	1	2	6	9	-
2.1.1	2	1	7	8	-
2.1.1	2	2	7	9	-
3	2	2	7	9	No elements in common to both strings.

Algorithm: Checks to see if a string of integers $x_1, x_2, \ldots, x_n$ contains any duplicate integers.

1. Input string $x_1, \ldots, x_n$.
2. **For** $i = 1$ to $n - 1$ **do**
   2.1 **For** $j = i + 1$ to n **do**
      2.1.1 **If** $x_i = x_j$ **then**
         2.1.1.1 Return "There is duplicate integer in the string."
3. Return "There are no duplicate integers in the string."

Example 1.8

Use the above algorithm to check if the string 98767 contains any duplicate integers.

The input is $x_1 = 9$, $x_2 = 8$, $x_3 = 7$, $x_4 = 6$, $x_5 = 7$, and $n = 5$.

Step	i	j	x_i	x_j	Output
2.1.1	1	2	9	8	-
2.1.1	1	3	9	7	-
2.1.1	1	4	9	6	-
2.1.1	1	5	9	7	-
2.1.1	2	3	8	7	-
2.1.1	2	4	8	6	-
2.1.1	2	5	8	7	-
2.1.1	3	4	7	6	-
2.1.1	3	5	7	7	-
2.1.1.1	3	5	7	7	There is a duplicate in the string.

Algorithm: Determines if a string of characters $x_1, x_2, \ldots, x_n$ consists entirely of integers.

1. Input string $x_1, x_2, \ldots, x_n$ and string length n.
2. $i \longleftarrow 1$
3. *noninteger_detected* $\longleftarrow$ *false*
4. **Repeat**
   4.1. **If** x_i is not an integer **then**
      4.1.1. *noninteger_detected* $\longleftarrow$ *true*
   4.2. $i \longleftarrow i + 1$
   until *noninteger_detected* $=$ *true* or $i = n + 1$
5. **If** *noninteger_detected* $=$ *true* **then**
   5.1 Output "String contains non-integer characters."
   else
   5.2 Output "String consists entirely of integers."

Example 1.9

Use the above algorithm to determine if the string $68c5$ consists entirely of integers or if a non-integer is contained in the string.

The input it $x_1 = 6$, $x_2 = 8$, $x_3 = c$, and $x_4 = 5$, and string length $n = 4$.

Step	nondigit_detected	i	x_i	Output
2	-	1	-	-
3	false	1	-	-
4.1	false	1	6	-
4.2	false	2	6	-
4.1	false	2	8	-
4.2	false	3	8	-
4.1	false	3	c	-
4.1.1	true	3	c	-
4.2	true	4	c	-
5.1	true	4	c	String contains non-integer characters.

1.5 PROBLEMS

Question 1.1 *Give a description of what the following algorithm does. What is the output of this algorithm for the input days = 3, hours = 7, minutes = 51, seconds = 27?*

1. Input *days, hours, minutes, seconds.*
2. *answer* $\longleftarrow$ *seconds* $+ (minutes \times 60) + (hours \times 60 \times 60) + (days \times 24 \times 60 \times 60)$
3. Output *answer.*

Question 1.2 *Trace the below algorithm for the following pairs of numbers.*
  (a) $x = 6$ *and* $y = 12$ *(b)* $x = 15$ *and* $y = 9$ *(c)* $x = 7$ *and* $y = 7$

1. Input *x* and *y.*
2. **If** $x > y$ **then**
    2.1 *max* $\longleftarrow$ *x*
    else
    2.2 *max* $\longleftarrow$ *y*
3. Output *max.*

Question 1.3 *Trace the below algorithm for the following pairs of numbers.*
  (a) price $= 74.36$, paid $= 82.50$ *(b)* price $= 75.00$, paid $= 55.50$ *(c)* price $= 48.36$, paid $= 54.92$

1. Input *price* and *paid*.
2. **If** *paid* < *price* **then**
      2.1 Output "Did not pay enough."
     else
      2.2 *change* ⟵ *paid* − *price*
3. Output "Change *change*."

Question 1.4 *Trace the below algorithm for the following pairs of numbers.*
(a) $x = 1$ (b) $x = -3$ (c) $x = 7$

1. Input x.
2. $x \longleftarrow -x$
3. **If** $x = -1$ **then**
      3.1. *answer* ⟵ $x + 5$
     else
      3.2. *answer* ⟵ $x - 5$
4. Output *answer*.

Question 1.5 *Trace the below algorithm for the following pairs of numbers.*
(a) $x = -6$ (b) $x = 2$ (c) $x = 8$

1. Input x.
2. **If** $x < -5$ **then**
      2.1. *answer* ⟵ $2x + 3$
     else
      2.2 **If** $x > 5$ **then**
         2.2.1 *answer* ⟵ $-2x + 3$
       else
         2.2.2 *answer* ⟵ -7
3. Output *answer*.

Question 1.6 *Trace the below algorithm for the following pairs of numbers.*
(a) $x = 2$ *and* $n = 4$ (b) $x = 3$ *and* $n = 5$ (c) $x = 5$ *and* $n = 6$

1. Input x and n.
2. *answer* ⟵ x
3. **For** $i = 1$ to $n - 1$ **do**
      3.1. *answer* ⟵ *answer* $\times x$
4. Output *answer*.

Question 1.7 *Trace the below algorithm for the following numbers.*
(a) $n = 3$ (b) $n = 5$ (c) $n = 7$

1. Input positive integer n.
2. *fac* ⟵ n
3. **While** $n > 1$ **do**
      3.1 $n \longleftarrow n - 1$
      3.2 *fac* ⟵ *fac* $\times n$
4. Output *fac*.

Question 1.8 *Trace the below algorithm for the following numbers. Then compare it to the algorithm in the last question.*

(a) $n = 3$ (b) $n = 5$ (c) $n = 7$

1. Input positive integer n.
2. *fac* ⟵ 1
3. **For** $i = 1$ to n **do**
       3.1 *fac* ⟵ *fac* × i
4. Output *fac*.

Question 1.9 *Trace the following algorithm for* $n = 8, 490, 725, 727, 154, 368, 726, 402, 945$.

1. Input a positive integer n.
2. d ⟵ number of digits of n
3. **While** $d > 1$ **do**
       3.1. n ⟵ sum of digits of n
       3.2 d ⟵ number of digits of n
4. Output n.

Question 1.10 *Trace the following algorithm for*

(a) *string of integers 274390 to see if the integer $s = 9$ is contained in the string,*

(b) *string of integers 730285 to see if the integer $s = 2$ is contained in the string,*

(c) *string of integers 983650362 to see if the integer $s = 5$ is contained in the string.*

1. Input string $x_1, \ldots, x_n$, n and s.
2. **For** $i = 1$ to n **do**
       2.1. **If** $x_i = s$ **then**
          2.1.1. Return "String contains s."
3. Return "String does not contain s."

Question 1.11 *Trace the below algorithm for*

(a) *strings 2109 and 4071 to see if the integer $s = 0$ is contained in either one,*

(b) *strings 593 and 721 to see if the integer $s = 2$ is contained in either one,*

(c) *strings 397 and 443 to see if the integer $s = 6$ is contained in either one.*

1. Input strings $x_1, \ldots, x_n$ and $y_1, \ldots, y_n$, n and s.
2. **For** $i = 1$ to n **do**
       2.1. **If** $x_i = s$ **then**
          2.1.1. Return "String one contains integer s."
3. **For** $i = 1$ to n **do**
       3.1. **If** $y_i = s$ **then**
          3.1.1. Return "String two contains integer s."
4. Return "Neither string contains integer s."

Question 1.12 *Trace the below algorithm for*

(a) *strings 273 and 439,*   (b) *strings 593 and 725,*   (c) *strings 123 and 987.*

> 1. Input strings $x_1, \ldots, x_n$ and $y_1, \ldots, y_n$.
> 2. **For** $i = 1$ to n **do**
>   2.1 **For** $j = 1$ to n **do**
>   2.1.1 **If** $x_i = y_j$ **then**
>   2.1.1.1 Return "There is an element common to both strings."
> 3. Return "There are no elements common to both strings."

Question 1.13 *Trace the below algorithm for*

(a) *string 63524,*   (b) *string 47372916,*   (c) *string 563732.*

> 1. Input string $x_1, \ldots, x_n$.
> 2. **For** $i = 1$ to $n - 1$ **do**
>   2.1 **For** $j = i + 1$ to n **do**
>   2.1.1 **If** $x_i = x_j$ **then**
>   2.1.1.1 Return "There is a duplicate integer in the string."
> 3. Return "There are no duplicate integers in the string."

Question 1.14 *Trace the below algorithm for*

(a) $n = 3$,   (b) $n = 4$,   (c) $n = 6$.

> 1. Input n. (n a positive number.)
> 2. *sum* ⟵ 0
> 3. **For** $i = 1$ to n **do**
>   3.1. *sum* ⟵ $sum + i^2$
> 4. Output *sum*.

Question 1.15 *Trace the below algorithm for*

(a) $n = 12$,   (b) $n = 90$,   (c) $n = 80$.

> 1. Input a non-negative integer n.
> 2. *order* ⟵ 0
> 3. **While** n is even **do**
>   3.1. $n \leftarrow \frac{n}{2}$
>   3.2. *order* ⟵ *order* + 1
> 4. Output *order*.

Question 1.16 *Give a description of what the following algorithm does. Then trace this algorithm for*

(a) *string* 5, 7, 4, 6, 2, 8,   (b) *string* 3, 5, 7, 2, 4, 6,   (c) *string* 7, 2, 5, 0, 9, 3, 1.

> 1. Input list of numbers $x_1, x_2, \ldots, x_n$.
> 2. *min* ⟵ x_1, *position* ⟵ 1
> 3. **For** $i = 2$ to n **do**
>   3.1. **If** $x_i < min$ **then**
>   3.1.1 *min* ⟵ x_i
>   3.1.2 *position* ⟵ i
> 4. Output *min, position*.

Question 1.17 *Trace the below algorithm for*
  (a) the string 04m3s8,   *(b) the string 732r9,*   *(c) the string 890.*

1. Input string $x_1, x_2, \ldots, x_n$.
2. $i \longleftarrow 1$
3. *noninteger_detected* $\longleftarrow$ *false*
4. **Repeat**
   4.1. **If** x_i is not an integer **then**
       4.1.1. *noninteger_detected* $\longleftarrow$ *true*
   4.2. $i \longleftarrow i + 1$
   until *noninteger_detected* $=$ *true* or $i = n + 1$
5. **If** *noninteger_detected* $=$ *true* **then**
   5.1 Output "String contains non-integer characters."
   else
   5.2 Output "String consists entirely of integers."

Question 1.18 *Rewrite the above algorithm using returns.*

CHAPTER 2

Number Representations

Binary, or base-two, numbers are extremely important in computer science because they are used in designing computers. Computer scientists also use octal and hexadecimal numbers, which are closely related to binary numbers. Therefore computer science majors need to have a good understanding of all these number systems and be able to convert between them and decimal, or base-ten, numbers.

2.1 WHOLE NUMBERS

Decimal numbers are often called base-ten numbers. Decimal numbers can be written in **expanded form** as follows,

$$8 = 8 \times 10^0,$$
$$72 = 7 \times 10^1 + 2 \times 10^0,$$
$$401 = 4 \times 10^2 + 0 \times 10^1 + 1 \times 10^0,$$
$$8925 = 8 \times 10^3 + 9 \times 10^2 + 2 \times 10^1 + 5 \times 10^0.$$

It is important to remember that $10^0 = 1$ and to remember the order of operations. The exponents are done first, then all the multiplications are done from left to right, then all the additions are done from left to right. In the above example the number 10 is called the **base**. Notice that when we use a base 10 the **coefficients** of the powers of ten, that is, the numbers that are in front of the powers of ten, range from 0 to 9. These coefficients are also call **digits**.

We can write numbers with different bases. For example, instead of using a 10 as we did above we can use a 2. Numbers that use a two as a base are called either base-two numbers or **binary** numbers. The coefficients of the powers of two range from 0 to 1. In other words, binary numbers have the digits 1 and 0. Binary numbers are very important in computer science because these are the numbers that computers actually use. Computers use electricity, and either there is an electrical current or there isn't. This means computers can only "understand" two states; is there an electrical current present or is there no electrical current[1] present? These two states are often called called 1 and 0. So any numbers computers use can only have two coefficients, or digits. Numbers that only have two digits are binary

---

[1]This is actually a bit of a simplification, there is either a high level of current or a very low level of residual current.

numbers and have a base of 2. Binary numbers can be written in expanded form as follows

$$1 = 1 \times 2^0,$$
$$10 = 1 \times 2^1 + 0 \times 2^0,$$
$$101 = 1 \times 2^2 + 0 \times 2^1 + 1 \times 2^0,$$
$$1011 = 1 \times 2^3 + 0 \times 2^2 + 1 \times 2^1 + 1 \times 2^0.$$

Again, remember that $2^0 = 1$ and remember the order of operations. We always need to know what base a number is written in. Sometimes it is obvious what base the numbers should have. Sometimes it isn't. When it is not clear what base a number is written in sometimes a subscript is used. A subscript of 2 is used to indicate a binary number and a subscript of 10 is used to indicate a decimal number. Here

$$110_2 \text{ is a base-two, or binary, number,}$$
$$401_{10} \text{ is a base-ten, or decimal, number,}$$
$$1011_2 \text{ is a base-two, or binary, number,}$$
$$7203_{10} \text{ is a base-ten, or decimal, number.}$$

It is easy to convert from binary numbers to decimal numbers. As we said above, the digits for binary numbers are simply 0 and 1. It should be clear that 0 in base-two, or 0_2 is exactly the same as 0 in base-ten, or 0_{10}. Similarly, $1_2 = 1_{10}$. Therefore, in the below example there is no need to specify which base the digits are in since they are the same in both bases.

Example 2.1

- Convert 110_2 to a decimal number.

$$110_2 = 1 \times \underbrace{2^2}_{=4_{10}} + 1 \times \underbrace{2^1}_{=2_{10}} + 0 \times \underbrace{2^0}_{=1_{10}}$$
$$= \underbrace{1 \times 4_{10}}_{=4_{10}} + \underbrace{1 \times 2_{10}}_{=2_{10}} + \underbrace{0 \times 1_{10}}_{=0_{10}}$$
$$= 4_{10} + 2_{10} + 0_{10}$$
$$= 6_{10}$$

- Convert 1011_2 to a decimal number.

$$1011_2 = 1 \times 2^3 + 0 \times 2^2 + 1 \times 2^1 + 1 \times 2^0$$
$$= 1 \times 8_{10} + 0 \times 4_{10} + 1 \times 2_{10} + 1 \times 1_{10}$$
$$= 8_{10} + 0_{10} + 2_{10} + 1_{10}$$
$$= 11_{10}$$

Very quickly binary numbers become difficult for us to understand. For example, consider the number

$$10001111101010100011100011101$$

and the number

$$10001111101110100011100011101.$$

Just by looking can you easily tell which of these two numbers was larger? Even though computers use binary numbers it is difficult for us to use them. So there are two other basis that are very important in computer science, base-eight and base-sixteen.

For base-eight numbers instead of using a 10 or a 2 we use an 8 as the base. Base-eight numbers are also called **octal** numbers. The coefficients of octal numbers range from 0 to 7. This means that octal numbers use the digits $0, 1, 2, 3, 4, 5, 6, 7$. Sometimes we know from context that a number is octal, but sometimes it is not clear. In this case we often use a subscript of 8 to indicate the number is an octal number. Octal numbers can be written in expanded form,

$$5_8 = 5 \times 8^0,$$
$$70_8 = 7 \times 8^1 + 0 \times 8^0,$$
$$372_8 = 3 \times 8^2 + 7 \times 8^1 + 2 \times 8^0,$$
$$6504_8 = 6 \times 8^3 + 5 \times 8^2 + 0 \times 8^1 + 4 \times 2^0.$$

Again, it is easy to convert from octal numbers to decimal numbers. It should be clear that 0 in base-eight, or 0_8 is exactly the same as 0 in base-ten, or 0_{10}. Similarly, $1_8 = 1_{10}$ and so on up to $7_8 = 7_{10}$. Therefore, in the below example there is no need to specify which base the digits 0 through 7 are since they are the same in both bases.

Example 2.2

- Convert 70_8 to a decimal number.

$$
\begin{aligned}
70_8 &= 7 \times 8^1 + 0 \times 8^0 \\
&= 7 \times 8_{10} + 0 \times 1_{10} \\
&= 56_{10} + 0_{10} \\
&= 56_{10}
\end{aligned}
$$

- Convert 6504_8 to a decimal number.

$$
\begin{aligned}
6504_8 &= 6 \times 8^3 + 5 \times 8^2 + 0 \times 8^1 + 4 \times 8^0 \\
&= 6 \times 512_{10} + 5 \times 64_{10} + 0 \times 8_{10} + 4 \times 1_{10} \\
&= 3072_{10} + 320_{10} + 0_{10} + 4_{10} \\
&= 3396_{10}
\end{aligned}
$$

For base-sixteen numbers we use 16 as the base. Base-sixteen numbers are also called **hexadecimal** numbers. Like before, we want the coefficients of the powers of 16 to range from 0 to 15. The problem is that the numbers 10 through 15 each have two decimal digits in them. Using these would be very inconvenient. Therefore we use the capital letters A, B, C, D, E, and F instead. The table below summarizes the correspondence between the decimal numbers and the digits used for hexadecimal numbers.

This means hexadecimal numbers look something like $3B7$, $A05C$, $E2$, or D. Hexadecimal numbers can be written in expanded form

$$D_{16} = D \times 16^0,$$
$$E2_{16} = E \times 16^1 + 2 \times 16^0,$$
$$3B7_{16} = 3 \times 16^2 + B \times 16^1 + 7 \times 16^0,$$
$$A05C_{16} = A \times 16^3 + 0 \times 16^2 + 5 \times 16^1 + C \times 16^0.$$

Decimal Numbers	Hexadecimal Digits
0_{10}	0
1_{10}	1
2_{10}	2
3_{10}	3
4_{10}	4
5_{10}	5
6_{10}	6
7_{10}	7
8_{10}	8
9_{10}	9
10_{10}	A
11_{10}	B
12_{10}	C
13_{10}	D
14_{10}	E
15_{10}	F

Converting hexadecimal numbers to decimal numbers works just like before, except that we need to change the hexadecimal digits to decimal digits using the table. As with the binary and octal case it is obvious that $0_{16} = 0_{10}$ and so on up to $9_{16} = 9_{10}$. Therefore, as before we make no effort to specify which base the digits 0 through 9 are since they are the same in both bases. However, when we convert from the digits A through F we go ahead and specify the base. That is, we write A as 10_{10} and so on up to writing F as 15_{10}.

Example 2.3

- Convert $3B7_{16}$ to a decimal number.

$$3B7_{16} = 3 \times 16^2 + \underbrace{B}_{=11_{10}} \times 16^1 + 7 \times 16^0$$

$$= 3 \times 256_{10} + 11_{10} \times 16_{10} + 7 \times 1_{10}$$

$$= 768_{10} + 176_{10} + 7_{10}$$

$$= 951_{10}$$

- Convert $A05C_{16}$ to a decimal number.

$$A05C_{16} = \underbrace{A}_{=10_{10}} \times 16^3 + 0 \times 16^2 + 5 \times 16^1 + \underbrace{C}_{=12_{10}} \times 16^0$$

$$= 10_{10} \times 4096_{10} + 0 \times 256_{10} + 5 \times 16_{10} + 12_{10} \times 1_{10}$$

$$= 40960_{10} + 0_{10} + 80_{10} + 12_{10}$$

$$= 41052_{10}$$

2.2 FRACTIONAL NUMBERS

We all know what a decimal point is. That is, we know what numbers like 7.4 or 50.69 or 382.207 mean. Often numbers with a decimal point are called decimal numbers, but we are already using that word to describe base-ten numbers so here we will call them **fractional** numbers. Fractional numbers have a point and some digits after the point. Fractional base-ten numbers can be written in an expanded form just like before,

$$50.69_{10} = 5 \times 10^1 + 0 \times 10^0 + 6 \times 10^{-1} + 9 \times 10^{-2},$$
$$382.207_{10} = 3 \times 10^2 + 8 \times 10^1 + 2 \times 10^0 + 2 \times 10^{-1} + 0 \times 10^{-2} + 7 \times 10^{-3}.$$

Here we need to remember how to work with negative exponents,

$$10^{-1} = \frac{1}{10^1} = \frac{1}{10} = 0.1_{10},$$
$$10^{-2} = \frac{1}{10^2} = \frac{1}{100} = 0.01_{10},$$
$$10^{-3} = \frac{1}{10^3} = \frac{1}{1000} = 0.001_{10},$$

and so on. Thus, if we take a closer look we can see what is really happening,

$$50.69_{10} = 5 \times 10 + 0 \times 1 + 6 \times 0.1 + 9 \times 0.01$$
$$= 5 \times 10^1 + 0 \times 10^0 + 6 \times 10^{-1} + 9 \times 10^{-2}.$$

Fractional binary numbers can be written in expanded form similarly,

$$101.01_2 = 1 \times 2^2 + 0 \times 2^1 + 1 \times 2^0 + 0 \times 2^{-1} + 1 \times 2^{-2},$$
$$1101.101_2 = 1 \times 2^3 + 1 \times 2^2 + 0 \times 2^1 + 1 \times 2^0 + 1 \times 2^{-1} + 0 \times 2^{-2} + 1 \times 2^{-3}.$$

To convert fractional binary numbers to fractional decimal numbers we need to know the negative powers of two,

$$2^{-1} = \frac{1}{2^1} = \frac{1}{2} = 0.5_{10},$$
$$2^{-2} = \frac{1}{2^2} = \frac{1}{4} = 0.25_{10},$$
$$2^{-3} = \frac{1}{2^3} = \frac{1}{8} = 0.125_{10}.$$

Example 2.4

- Convert 10.11_2 to a decimal number.

$$10.11_2 = 1 \times 2^1 + 0 \times 2^0 + 1 \times 2^{-1} + 1 \times 2^{-2}$$
$$= 1 \times 2_{10} + 0 \times 1_{10} + 1 \times 0.5_{10} + 1 \times 0.25_{10}$$
$$= 2_{10} + 0_{10} + 0.5_{10} + 0.25_{10}$$
$$= 2.75_{10}$$

- Convert 101.01_2 to a decimal number.

$$101.01_2 = 1 \times 2^2 + 0 \times 2^1 + 1 \times 2^0 + 0 \times 2^{-1} + 1 \times 2^{-2}$$
$$= 1 \times 4_{10} + 0 \times 2_{10} + 1 \times 1_{10} + 0 \times 0.5_{10} + 1 \times 0.25_{10}$$
$$= 4_{10} + 0_{10} + 1_{10} + 0_{10} + 0.25_{10}$$
$$= 5.25_{10}$$

Fractional octal numbers of course work exactly the same way,

$$731.42_8 = 7 \times 8^2 + 3 \times 8^1 + 1 \times 8^0 + 4 \times 8^{-1} + 2 \times 8^{-2},$$
$$1605.734_8 = 1 \times 8^3 + 6 \times 8^2 + 0 \times 8^1 + 5 \times 8^0 + 7 \times 8^{-1} + 3 \times 8^{-2} + 4 \times 8^{-3}.$$

In order to convert fractional octal numbers to fractional decimal numbers we need to know the negative powers of eight,

$$8^{-1} = \frac{1}{8^1} = \frac{1}{8} = 0.125_{10},$$
$$8^{-2} = \frac{1}{8^2} = \frac{1}{64} = 0.015625_{10},$$
$$8^{-3} = \frac{1}{8^3} = \frac{1}{512} = 0.001953125_{10}.$$

This starts to get messy, but for the questions in this book use all digits after the point. In other words, do not round.

Example 2.5

- Convert 3.7_8 to a decimal number.

$$3.7_8 = 3 \times 8^0 + 7 \times 8^{-1}$$
$$= 3 \times 1_{10} + 7 \times 0.125_{10}$$
$$= 3_{10} + 0.875_{10}$$
$$= 3.875_{10}$$

- Convert 731.42_8 to a decimal number.

$$731.42_8 = 7 \times 8^2 + 3 \times 8^1 + 1 \times 8^0 + 4 \times 8^{-1} + 2 \times 8^{-2}$$
$$= 7 \times 64_{10} + 3 \times 8_{10} + 1 \times 1_{10} + 4 \times 0.125_{10} + 2 \times 0.015625_{10}$$
$$= 448_{10} + 24_{10} + 1_{10} + 0.5_{10} + 0.03125_{10}$$
$$= 473.53125_{10}$$

Fractional hexadecimal numbers behave similarly, they just use the digits 0 through F. Fractional hexadecimal numbers can be written in expanded form,

$$8A.41_{16} = 8 \times 16^1 + A \times 16^0 + 4 \times 16^{-1} + 1 \times 16^{-2},$$
$$A0D.4EB_{16} = A \times 16^2 + 0 \times 16^1 + D \times 16^0 + 4 \times 16^{-1} + E \times 16^{-2} + B \times 16^{-3}.$$

In order to convert fractional hexadecimal numbers to fractional decimal numbers we need to know the negative powers of 16,

$$16^{-1} = \frac{1}{16^1} = \frac{1}{16} = 0.0625_{10},$$
$$16^{-2} = \frac{1}{16^2} = \frac{1}{256} = 0.00390625_{10},$$
$$16^{-3} = \frac{1}{16^3} = \frac{1}{4096} = 0.000244140625_{10}.$$

Example 2.6

- Convert $C.3_{16}$ to a decimal number.

$$C.3_{16} = C \times 16^0 + 3 \times 16^{-1}$$
$$= 12_{10} \times 1_{16} + 3 \times 0.0625_{10}$$
$$= 12_{10} + 0.1875_{10}$$
$$= 12.1875_{10}$$

- Convert $8A.41_{16}$ to a decimal number.

$$8A.41_{16} = 8 \times 16^1 + A \times 16^0 + 4 \times 16^{-1} + 1 \times 16^{-2}$$
$$= 8 \times 16_{10} + 10_{10} \times 1_{10} + 4 \times 0.0625_{10} + 1 \times 0.00390625_{10}$$
$$= 128_{10} + 10_{10} + 0.25_{10} + 0.00390625_{10}$$
$$= 138.25390625_{10}$$

2.3 THE RELATIONSHIP BETWEEN BINARY, OCTAL, AND HEXADECIMAL NUMBERS

Earlier we said that very quickly binary numbers become difficult for us to understand and asked you which of the numbers, 1000111110101010011100011101 or 1000111110111010011100011101, was larger just by looking at them. Probably you would have to look a while before figuring it out. In a lot of ways binary numbers are rather difficult to work with so computer scientists would prefer to be able to work with other number systems that are easier for a human to understand. It turns out that either octal or hexadecimal numbers are the best number systems for computer scientists to work with. The reason that these two number systems are convenient is how easy it is to convert binary numbers to either octal numbers or hexadecimal numbers and then back again. Because of this ease, computer scientists prefer these systems over the decimal system. With enough practice they become very easy to use.

It should be easy for you to verify the correspondences shown in the table below. To convert from octal numbers to binary numbers we just replace each octal digit with its corresponding three digit binary number from the table.

Binary Numbers	Octal Digits
000	0
001	1
010	2
011	3
100	4
101	5
110	6
111	7

Example 2.7

Converting octal numbers to binary numbers.

- Convert the octal number 34_8 to a binary number. Notice we can drop the leading zero from the binary number without changing the value of the number.

$$34_8 = \underbrace{3}_{011}\ \underbrace{4}_{100}$$
$$= 011100_2$$
$$= 11100_2$$

- Convert the octal number 17045_8 to a binary number.

$$17043_8 = \underbrace{1}_{001}\ \underbrace{7}_{111}\ \underbrace{0}_{000}\ \underbrace{4}_{100}\ \underbrace{3}_{011}$$
$$= 001111000100011_2$$
$$= 1111000100011_2$$

- Convert the octal number 35.204_8 to a binary number. Notice we can also drop the final zeros **after the point** without changing the number. We can of course not drop final zeros before the point. That would change the value of the number.

$$35.204_8 = \underbrace{3}_{011}\ \underbrace{5}_{101}\ .\ \underbrace{2}_{010}\ \underbrace{0}_{000}\ \underbrace{4}_{100}$$
$$= 011101\ .\ 010000100_2$$
$$= 11101\ .\ 0100001_2$$

Converting from binary numbers to octal numbers is equally easy. But pay close attention, it is easy to get confused.

- For a **whole number** we group the **binary digits (before the point if there is one) in sets of three from right to left**, adding leading zeros if necessary, and then replace the three digit binary numbers with their corresponding octal digits from the table.
- For **fractional numbers** we group the **binary digits after the point in sets of three from left to right**, adding zeros at the end if necessary. Then we replace the three digit binary numbers with their corresponding octal digits.

Example 2.8

Converting binary numbers to octal numbers. The second example illustrates filling in zeros at the start of the binary number and the third example illustrates filling in zeros at both the beginning and the end of the number.

- Convert 101110001010_2 to an octal number.

$$101110001010_2 = \underbrace{101}_{5}\ \underbrace{110}_{6}\ \underbrace{001}_{1}\ \underbrace{010}_{2}$$
$$= 5612_8$$

- Convert 1001101101001_2 to an octal number.

$$1001101101001_2 = \underbrace{1}_{\substack{=001 \\ 1}} \quad \underbrace{001}_{1} \; \underbrace{101}_{5} \; \underbrace{101}_{5} \; \underbrace{001}_{1}$$

$$= 11551_8$$

- Convert 10100111.1001101_2 to an octal number.

$$10100111.1001101_2 = \underbrace{10}_{\substack{=010 \\ 2}} \; \underbrace{100}_{4} \; \underbrace{111}_{7} . \underbrace{100}_{4} \; \underbrace{110}_{6} \; \underbrace{1}_{\substack{=100 \\ 4}}$$

$$= 247.464_8$$

Example 2.9

Which number, $1000111110101010011100011101_2$ or $1000111110111010011100011101_2$, is larger? We convert the first binary number into an octal number.

$$1000111110101010011100011101_2 = \underbrace{1}_{\substack{=001 \\ 1}} \; \underbrace{000}_{0} \; \underbrace{111}_{7} \; \underbrace{110}_{6} \; \underbrace{101}_{5} \; \underbrace{010}_{2} \; \underbrace{011}_{3} \; \underbrace{100}_{4} \; \underbrace{011}_{3} \; \underbrace{101}_{5}$$

$$= 1076523435_8$$

Next we convert the second binary number into an octal number.

$$1000111110111010011100011101_2 = \underbrace{1}_{\substack{=001 \\ 1}} \; \underbrace{000}_{0} \; \underbrace{111}_{7} \; \underbrace{110}_{6} \; \underbrace{111}_{7} \; \underbrace{010}_{2} \; \underbrace{011}_{3} \; \underbrace{100}_{4} \; \underbrace{011}_{3} \; \underbrace{101}_{5}$$

$$= 1076723435_8$$

Now we compare the two numbers;

$$1076523435_8,$$

$$1076723435_8.$$

This is a much easier job than comparing the two original binary numbers. But it could be made easier yet by using hexadecimal numbers.

This table shows the correspondences that are necessary to convert between binary numbers and hexadecimal numbers. It should be easy for you to verify the relations in the table below. To convert hexadecimal numbers to binary numbers we just replace each hexadecimal digit with the corresponding binary number.

Binary Numbers	Hexadecimal Digits
0000	0
0001	1
0010	2
0011	3
0100	4
0101	5
0110	6
0111	7
1000	8
1001	9
1010	A
1011	B
1100	C
1101	D
1110	E
1111	F

Example 2.10

Converting hexadecimal numbers to binary numbers.

- Convert the hexadecimal number $E2_{16}$ to a binary number.

$$E2_{16} = \underbrace{E}_{1110} \; \underbrace{2}_{0010}$$
$$= 11100010_2$$

- Convert the hexadecimal number $3F9A_{16}$ to a binary number.

$$3F9A_{16} = \underbrace{3}_{0011} \; \underbrace{F}_{1111} \; \underbrace{9}_{1001} \; \underbrace{A}_{1010}$$
$$= 11111110011010_2$$

- Convert the hexadecimal number $D06.0C_{16}$ to a binary number.

$$D06.0C_{16} = \underbrace{D}_{1101} \; \underbrace{0}_{0000} \; \underbrace{6}_{0110} . \underbrace{0}_{0000} \; \underbrace{C}_{1100}$$
$$= 110100000110.000011_2$$

Converting from binary numbers to hexadecimal numbers is similar to converting from binary to octal numbers. But again, pay close attention to the directions, it is easy to get confused.

- For a **whole number** we group the **binary digits (before the point if there is one) in sets of four from right to left**, adding leading zeros if necessary, and then replace the four digit binary numbers with their corresponding hexadecimal digits from the table.
- For **fractional numbers** we group the **binary digits after the point in sets of four from left to right**, adding zeros at the end if necessary. Then we replace the four digit binary numbers with their corresponding hexadecimal digits.

So, when converting to octal numbers we use groups of three but when converting to hexadecimal numbers we use groups of four.

Example 2.11

Converting binary numbers to hexadecimal numbers. The second example illustrates filling in zeros at the start of the binary number and the third example illustrates filling in zeros at both the beginning and the end of the number.

- Convert 100111010110_2 to a hexadecimal number.

$$100111010110_2 = \underbrace{1001}_{9}\ \underbrace{1101}_{D}\ \underbrace{0110}_{6}$$

$$= 9D6_{16}$$

- Convert 1010011110_2 to a hexadecimal number.

$$1010011110_2 = \underbrace{\underbrace{10}_{= 0010}}_{2}\ \underbrace{1001}_{9}\ \underbrace{1110}_{E}$$

$$= 29E_{16}$$

- Convert 1101001011.001101_2 to a hexadecimal number.

$$1101001011.001101_2 = \underbrace{\underbrace{11}_{= 0011}}_{3}\ \underbrace{0100}_{4}\ \underbrace{1011}_{B}.\underbrace{0011}_{3}\ \underbrace{\underbrace{01}_{= 0100}}_{4}$$

$$= 34B.34_{16}$$

Example 2.12

Which number, $10001111101010100011000011101_2$ or $10001111101110100011000011101_2$, is larger? We convert the first binary number into a hexadecimal number.

$$10001111101010100011000011101_2 = \underbrace{1000}_{8}\ \underbrace{1111}_{F}\ \underbrace{1010}_{A}\ \underbrace{1010}_{A}\ \underbrace{0111}_{7}\ \underbrace{0001}_{1}\ \underbrace{1101}_{D}$$

$$= 8FAA71D_{16}$$

Next we convert the second binary number into a hexadecimal number.

$$10001111101110100011000011101_2 = \underbrace{1000}_{8}\ \underbrace{1111}_{F}\ \underbrace{1011}_{B}\ \underbrace{1010}_{A}\ \underbrace{0111}_{7}\ \underbrace{0001}_{1}\ \underbrace{1101}_{D}$$

$$= 8FBA71D_{16}$$

Now we can compare these numbers;

$$8F\boldsymbol{A}A71D,$$
$$8F\boldsymbol{B}A71D.$$

This is a much easier job than comparing the two original binary numbers and even a little easier than comparing the octal numbers.

Converting between hexadecimal numbers and octal numbers requires that you use the binary number as an in-between step.

- To convert from a hexadecimal number to an octal number you first convert the hexadecimal number to a binary number and then convert the binary number to an octal number.
- To convert from an octal number to a hexadecimal number you first convert the octal number to a binary number and then convert the binary number to a hexadecimal number.

2.4 CONVERTING FROM DECIMAL NUMBERS

In Sections 2.1 and 2.2 we learned to convert binary numbers, octal numbers, and hexadecimal numbers into decimal numbers. In this section we will go in the other direction; we will convert decimal numbers into binary numbers, octal numbers, or hexadecimal numbers. To do this conversion we will use an algorithm. Since this class teaches the basic mathematics necessary for computer science majors, it is important that you get used to working with algorithms. We will begin with the algorithm for converting a whole decimal number to a binary number.

Algorithm: Converts a whole decimal number to a binary number.

1. Input n. (n is the whole decimal number.)
2. **Repeat**
   2.1 Output n mod **2**.
   2.2 $n \longleftarrow n$ div **2**
   until $n = 0$
3. Read outputs in reverse order.

In order to use this algorithm we need to understand what both n mod 2 and n div 2 mean. Both **mod** and **div** are words that define a mathematical operation. You have probably encountered the idea before under a different word.

- n div 2 = the **quotient** when n is divided by 2.
- n mod 2 = the **remainder** when n is divided by 2.

Let us consider the following long division, which hopefully you remember how to do.

$$
\begin{array}{r}
37 \\
2\overline{)75} \\
60 \\
\hline
15 \\
14 \\
\hline
1
\end{array}
$$

In this example the quotient is 37 and the remainder is 1. Thus we have

$$75 \text{ div } 2 = 37,$$
$$75 \text{ mod } 2 = 1.$$

It should be obvious to you that when finding n mod 2 for any number n the only possible remainders are 0 and 1 and so these are the only possible values that n mod 2 can take.

Example 2.13

Finding n div 2 and n mod 2 for $n = 8$ and $n = 9$.

- $n = 8$

$$
\begin{array}{r}
4 \\
2\,\overline{)\,8} \\
8 \\
\hline
0
\end{array}
$$

gives

8 div $2 = 4$

8 mod $2 = 0$

- $n = 9$

$$
\begin{array}{r}
4 \\
2\,\overline{)\,9} \\
8 \\
\hline
1
\end{array}
$$

gives

9 div $2 = 4$

9 mod $2 = 1$

Example 2.14

Finding n div 2 and n mod 2 for $n = 37$ and $n = 38$.

- $n = 37$

$$
\begin{array}{r}
18 \\
2\,\overline{)\,37} \\
20 \\
\hline
17 \\
16 \\
\hline
1
\end{array}
$$

gives

37 div $2 = 18$

37 mod $2 = 1$

We can also use our calculators, $37/2 = 18.5$. Here 37 div 2 is the whole number part of the answer, or 18. We get 37 mod 2 by taking the fractional part of the answer and multiplying it by 2, so we have $0.5 \times 2 = 1$.

- $n = 38$

$$
\begin{array}{r}
19 \\
2\,\overline{)\,38} \\
20 \\
\hline
18 \\
18 \\
\hline
0
\end{array}
$$

gives

38 div $2 = 19$

38 mod $2 = 0$

Using our calculators, $38/2 = 19$. Here 38 div 2 is the whole number part of the answer, or 19. We get 38 mod 2 by taking the fractional part of the answer and multiplying it by 2, but since the factional part is simply 0 we have $0 \times 2 = 0$.

In order to convert a decimal number into a binary number all we have to do is trace the algorithm and then use the output.

Example 2.15

Convert 11_{10} to a binary number. In order to do this we must follow the algorithm. As we follow the algorithm we will trace it in a table. Tracing the algorithm means we follow it step by step.

Step	n	Output
1.	11	-
2.1	11	$11 \bmod 2 = 1$
2.2	$5 \leftarrow 11 \text{ div } 2$	-
2.1	5	$5 \bmod 2 = 1$
2.2	$2 \leftarrow 5 \text{ div } 2$	-
2.1	2	$2 \bmod 2 = 0$
2.2	$1 \leftarrow 2 \text{ div } 2$	-
2.1	1	$1 \bmod 2 = 1$
2.2	$0 \leftarrow 1 \text{ div } 2$	-

Notice, when n became 0 the algorithm ended. Now we just have to read off the answer. As we can see from the algorithm we read the answer in reverse order, which means that we read the output numbers from the bottom to the top. Thus we have

$$11_{10} = 1011_2.$$

When doing this yourselves, you probably will not put as much detail in the trace table as we did. We just wanted to make this example very clear and easy to understand for you. We will do one more example to make sure you understand.

Example 2.16

Convert 37_{10} to a binary number. We will make a table that traces the algorithm.

Step	n	Output
1.	37	-
2.1	37	$37 \bmod 2 = 1$
2.2	$18 \leftarrow 37 \text{ div } 2$	-
2.1	18	$18 \bmod 2 = 0$
2.2	$9 \leftarrow 18 \text{ div } 2$	-
2.1	9	$9 \bmod 2 = 1$
2.2	$4 \leftarrow 9 \text{ div } 2$	-
2.1	4	$4 \bmod 2 = 0$
2.2	$2 \leftarrow 4 \text{ div } 2$	-
2.1	2	$2 \bmod 2 = 0$
2.2	$1 \leftarrow 2 \text{ div } 2$	-
2.1	1	$1 \bmod 2 = 1$
2.2	$0 \leftarrow 1 \text{ div } 2$	-

Again, when n became 0 the algorithm ended. Now we read the answer in reverse order, which means that we read the output numbers from the bottom to the top. Thus we have

$$37_{10} = 100101_2.$$

So far we have considered whole numbers. But we can convert fractional decimal numbers to binary numbers as well. Here is the algorithm to do this:

Algorithm: Converts a fractional decimal numbers to binary numbers.

1. Input n and d. (n is a fractional decimal number and d is digits.)
2. $i \longleftarrow 0$
3. **Repeat**
   3.1 $i \longleftarrow i + 1$
   3.2 $m \longleftarrow 2n$
   3.3 Output $\lfloor m \rfloor$.
   3.4 $n \longleftarrow \text{frac}(m)$
   until $n = 0$ or $i = d$
4. Read outputs in order. (Outputs follow the point.)

Notice that this algorithm requires us to decide how many digits after the point we want to have. In order to use this algorithm we have to understand what $\lfloor m \rfloor$ and $\text{frac}(m)$ mean. Fortunately, they are very easy to understand.

- $\lfloor m \rfloor$ = the whole number part of m

- $\text{frac}(m)$ = the fractional part of m

Usually $\lfloor m \rfloor$ is called the **floor** of m.

Example 2.17

Find $\lfloor m \rfloor$ and $\text{frac}(m)$ for $m = 93.91$.

$$\lfloor 93.91 \rfloor = 93$$
$$\text{frac}(93.91) = 0.91$$

Example 2.18

Find $\lfloor m \rfloor$ and $\text{frac}(m)$ for $m = 0.75$.

$$\lfloor 0.75 \rfloor = 0$$
$$\text{frac}(0.75) = 0.75$$

Example 2.19

Find $\lfloor m \rfloor$ and $\text{frac}(m)$ for $m = 16$.

$$\lfloor 16 \rfloor = 16$$
$$\text{frac}(16) = 0$$

Example 2.20

Convert 0.42_{10} to a binary number. Find five digits after the point. That is, $n = 0.42_{10}$ and $d = 5$.

Step	m	i	Output
1.	0.42	-	-
2.	0.42	0	-
3.1	0.42	$1 \leftarrow 0 + 1$	-
3.2	$0.84 \leftarrow 2(0.42)$	1	-
3.3	0.84	1	$\lfloor 0.84 \rfloor = 0$
3.4	$0.84 \leftarrow \text{frac}(0.84)$	1	-
3.1	0.84	$2 \leftarrow 1 + 1$	-
3.2	$1.68 \leftarrow 2(0.84)$	2	-
3.3	1.68	2	$\lfloor 1.68 \rfloor = 1$
3.4	$0.68 \leftarrow \text{frac}(1.68)$	2	-
3.1	0.68	$3 \leftarrow 2 + 1$	-
3.2	$1.36 \leftarrow 2(0.68)$	3	-
3.3	1.36	3	$\lfloor 1.36 \rfloor = 1$
3.4	$0.36 \leftarrow \text{frac}(1.36)$	3	-
3.1	0.36	$4 \leftarrow 3 + 1$	-
3.2	$0.72 \leftarrow 2(0.36)$	4	-
3.3	0.72	4	$\lfloor 0.72 \rfloor = 0$
3.4	$0.72 \leftarrow \text{frac}(0.72)$	4	-
3.1	0.72	$5 \leftarrow 4 + 1$	-
3.2	$1.44 \leftarrow 2(0.72)$	5	-
3.3	1.44	5	$\lfloor 1.44 \rfloor = 1$
3.4	$0.44 \leftarrow \text{frac}(1.44)$	5	-

Now we see that $i = 5$ and can stop. Notice that one does not check that if i is the same as $d = 5$ until after step 3.4, which allows us to generate the fifth digit in step 3.3. Now we read the outputs in order, placing them after the point, to give us

$$0.42_{10} \approx 0.01101_2.$$

Of course we could continue finding more digits. The answer we have obtained is not exact, it is just an approximation. To make this clear we will expand out 0.01101_2,

$$0.01101_2 = 0 \times 2^{-1} + 1 \times 2^{-2} + 1 \times 2^{-3} + 0 \times 2^{-4} + 1 \times 2^{-5}$$
$$= 0 \times 0.5_{10} + 1 \times 0.25_{10} + 1 \times 0.125_{10} + 0 \times 0.0625_{10} + 1 \times 0.03125_{10}$$
$$= 0.25_{10} + 0.125_{10} + 0.03125_{10}$$
$$= 0.40625_{10}.$$

So, to get a more accurate approximation for 0.42_{10}, we would need d to be larger.

In order to convert a mixed decimal number like 37.42_{10} into a binary number we would need to:

1. Split the number into the whole number part and the fractional number part.
2. Use the first algorithm on the whole number part.
3. Use the second algorithm on the fractional number part.
4. Join the two results together to get the final answer.[2]

Example 2.21

Convert 37.42_{10} to a binary number finding five digits after the point.

Finding 37.42_{10} as a binary number requires us to do the first algorithm on 37_{10} to get 100101_2 and the second algorithm on 0.42_{10} to get 0.01101_2, both of which we have already done. Joining these two answers together and we would have $37.42_{10} \approx 100101.01101_2$.

Now we consider converting a decimal number to an octal number. Notice that the only difference in the algorithms is that the two is replaced by an eight.

Algorithm: Converts a whole decimal number to an octal number.

1. Input n. (n is the whole decimal number.)
2. **Repeat**
   2.1 Output n mod **8**.
   2.2 $n \longleftarrow n$ div **8**
   until $n = 0$
3. Read outputs in reverse order.

Example 2.22

Convert 242_{10} to an octal number. We will not put as much detail into the table as before.

Step	n	Output
1.	242	-
2.1	242	2
2.2	30	-
2.1	30	6
2.2	3	-
2.1	3	3
2.2	0	-

Notice, when n became 0 the algorithm ended. Now we read the answer in reverse order to give us $242_{10} = 362_8$.

---

[2]The technical term for this "joining" is concatenation.

The algorithm to convert a fractional number to an octal number works the same as in the binary case, only the two is replaced by an eight.

Algorithm: Converts a fractional decimal number to an octal number.

1. Input n and d. (n is fractional decimal number and d is digits.)
2. $i \longleftarrow 0$
3. **Repeat**
   3.1 $i \longleftarrow i + 1$
   3.2 $m \longleftarrow 8n$
   3.3 Output $\lfloor m \rfloor$.
   3.4 $n \longleftarrow \text{frac}(m)$
   until $n = 0$ or $i = d$
4. Read outputs in order. (Outputs follow the point.)

Example 2.23

Convert 0.81_{10} to an octal number. Find six digits after the point. That is, $n = 0.81_{10}$ and $d = 6$.

Step	m	i	Output
1.	0.81	-	-
2.	0.81	0	-
3.1	0.81	1	-
3.2	6.48	1	-
3.3	6.48	1	6
3.4	0.48	1	-
3.1	0.48	2	-
3.2	3.84	2	-
3.3	3.84	2	3
3.4	0.84	2	-
3.1	0.84	3	-
3.2	6.72	3	-
3.3	6.72	3	6
3.4	0.72	3	-
3.1	0.72	4	-
3.2	5.76	4	-
3.3	5.76	4	5
3.4	0.76	4	-
3.1	0.76	5	-
3.2	6.08	5	-
3.3	6.08	5	6
3.4	0.08	5	-
3.1	0.08	6	-
3.2	0.64	6	-
3.3	0.64	6	0
3.4	0.64	6	-

Now we see that $i = 6$ and can stop. Thus we have $0.81_{10} \approx 0.636560_8$.

Example 2.24

Convert 242.81_{10} to an octal number using six digits after the point.

We have already used the first algorithm to find $242_{10} = 362_8$ and the second algorithm to find $0.81_{10} \approx 0.636560_8$. Joining these two numbers we now have $242.81_{10} \approx 362.636560_8$.

Now we consider converting a decimal number to a hexadecimal number. Again, the only difference in the algorithms is that we now use a sixteen.

Algorithm: Converts a whole decimal number to a hexadecimal number.

1. Input n. (n is the whole decimal number.)
2. **Repeat**
   2.1 Output n mod **16**.
   2.2 $n \longleftarrow n$ div **16**
   until $n = 0$
3. Read outputs in reverse order.

Example 2.25

Convert 3626_{10} to a hexadecimal number.

Step	n	Output
1.	3626	-
2.1	3636	$10_{10} = A_{16}$
2.2	226	-
2.1	226	$2_{10} = 2_{16}$
2.2	14	-
2.1	14	$14_{10} = E_{16}$
2.2	0	-

Now we read the outputs in reverse order to give us $3626_{10} = E2A_{16}$.

The algorithm to convert a fractional number to a hexadecimal number works the same as in the binary and octal cases, the only difference in the algorithms is that we now use a sixteen.

Algorithm: Converts a fractional decimal number to an hexadecimal number.

1. Input n and d. (n is fractional decimal number and d is digits.)
2. $i \longleftarrow 0$
3. **Repeat**
   3.1 $i \longleftarrow i + 1$
   3.2 $m \longleftarrow 16n$
   3.3 Output $\lfloor m \rfloor$.
   3.4 $n \longleftarrow \text{frac}(m)$
   until $n = 0$ or $i = d$
4. Read outputs in order. (Outputs follow the point.)

Example 2.26

Convert 0.63_{10} to a hexadecimal number. Find four digits after the point. That is, $n = 0.63_{10}$ and $d = 4$.

Step	m	i	Output
1.	0.63	-	-
2.	0.63	0	-
3.1	0.63	1	-
3.2	10.08	1	-
3.3	10.08	1	$10_{10} = A_{16}$
3.4	0.08	1	-
3.1	0.08	2	-
3.2	1.28	2	-
3.3	1.28	2	$1_{10} = 1_{16}$
3.4	0.28	2	-
3.1	0.28	3	-
3.2	4.48	3	-
3.3	4.48	3	$4_{10} = 4_{16}$
3.4	0.48	3	-
3.1	0.48	4	-
3.2	7.68	4	-
3.3	7.68	4	$7_{10} = 7_{16}$
3.4	0.68	4	-

Now we see that $i = 4$ and can stop. Reading the outputs in order we have $0.63_{10} \approx 0.A147_{16}$.

Example 2.27

Convert 3626.63_{10} to a hexadecimal number finding four digits after the point.

We have already used the first algorithm to find $3626_{10} = E2A_{16}$ and the second algorithm to find $0.63_{10} \approx 0.A147_{16}$. Joining these two numbers, we now have $3626.63_{10} \approx E2A.A147_{16}$.

2.5 PROBLEMS

Question 2.1 *Convert the following binary numbers to decimal numbers.*

(a) 101

(b) 110

(c) 111

(d) 1011

(e) 1101

(f) 1010

(g) 10010

(h) 10110

(i) 11111

Question 2.2 *Convert the following binary numbers to decimal numbers.*

(a) 0110 1001

(b) 1011 1100

(c) 0011 0111

(d) 1111 0101

(e) 1001 0110

(f) 0110 1000

(g) 1011 0110

(h) 0010 1011

(i) 1101 1101

Question 2.3 *Convert the following octal numbers to decimal numbers.*

(a) 27	(d) 103	(g) 4721
(b) 42	(e) 673	(h) 2451
(c) 35	(f) 360	(i) 7715

Question 2.4 *Convert the following octal numbers to decimal numbers.*

(a) 7254	(d) 17530	(g) 365310
(b) 1602	(e) 72501	(h) 772531
(c) 4640	(f) 11101	(i) 417524

Question 2.5 *Convert the following hexadecimal numbers to decimal numbers.*

(a) A3	(d) 3D2	(g) F001
(b) 9F	(e) 9A0	(h) 6D27
(c) 17	(f) AED	(i) 39CB

Question 2.6 *Convert the following hexadecimal numbers to decimal numbers.*

(a) 802	(d) 290B	(g) ABCDE
(b) 6A3	(e) 4C71	(h) 2E916
(c) F2E	(f) 1101	(i) 97CA0

Question 2.7 *Convert the following binary numbers to decimal numbers.*

(a) 0.11	(d) 0.110	(g) 0.0111
(b) 0.01	(e) 0.101	(h) 0.0011
(c) 0.10	(f) 0.011	(i) 0.1001

Question 2.8 *Convert the following binary numbers to decimal numbers.*

(a) 100.110	(d) 1101.1000	(g) 1111.1111
(b) 110.011	(e) 0110.0111	(h) 0001.0001
(c) 111.001	(f) 1011.1001	(i) 1100.0101

Question 2.9 *Convert the following octal numbers to decimal numbers.*

(a) 0.4	(d) 0.72	(g) 0.553
(b) 0.1	(e) 0.03	(h) 0.407
(c) 0.7	(f) 0.14	(i) 0.321

Question 2.10 *Convert the following octal numbers to decimal numbers.*

(a) 7.5	(d) 37.63	(g) 427.014
(b) 4.2	(e) 56.12	(h) 510.442
(c) 3.6	(f) 25.25	(i) 703.635

Question 2.11 *Convert the following hexadecimal numbers to decimal numbers.*

(a) 0.7	(d) 0.A8	(g) 0.C48
(b) 0.A	(e) 0.4C	(h) 0.379
(c) 0.3	(f) 0.82	(i) 0.ABC

Question 2.12 *Convert the following hexadecimal numbers to decimal numbers.*

(a) 9.3
(b) B.E
(c) 4.C

(d) AC.DC
(e) F2.39
(f) 5E.C5

(g) A04.BB8
(h) 7CF.9F0
(i) 101.101

Question 2.13 *Convert the following octal numbers to binary numbers.*

(a) 62
(b) 31
(c) 571

(d) 31.72
(e) 42.15
(f) 72.37

(g) 313.011
(h) 643.026
(i) 211.361

Question 2.14 *Convert the following hexadecimal numbers to binary numbers.*

(a) A0
(b) 3B
(c) 27

(d) F2.F2
(e) 5B.93
(f) 88.7C

(g) 214.3C5
(h) AD2.0BC
(i) 101.011

Question 2.15 *Convert the following binary numbers to octal numbers.*

(a) 10
(b) 110
(c) 11

(d) 110.111
(e) 011.010
(f) 101.001

(g) 11011.001001
(h) 100011.10111001
(i) 11110001.00011101

Question 2.16 *Convert the following binary numbers to hexadecimal numbers.*

(a) 1101
(b) 10
(c) 101

(d) 1110.1101
(e) 1011.1010
(f) 0101.0011

(g) 111001011.0011011
(h) 1011101.01011010101
(i) 10111100101.100111

Question 2.17 *Convert the following decimal numbers to binary numbers.*

(a) 7
(b) 3
(c) 6

(d) 28
(e) 39
(f) 79

(g) 381
(h) 643
(i) 569

Question 2.18 *Convert the following decimal numbers to octal numbers.*

(a) 78
(b) 93
(c) 52

(d) 295
(e) 944
(f) 641

(g) 4862
(h) 5078
(i) 1532

Question 2.19 *Convert the following decimal numbers to hexadecimal numbers.*

(a) 34
(b) 23
(c) 536

(d) 285
(e) 1897
(f) 8321

(g) 90867
(h) 42778
(i) 28714

Question 2.20 *Convert the following decimal numbers to binary numbers. Stop after you have obtained six digits.*

(a) 0.5
(b) 0.4
(c) 0.8

(d) 0.32
(e) 0.81
(f) 0.77

(g) 0.239
(h) 0.552
(i) 0.798

Question 2.21 *Convert the following decimal numbers to octal numbers. Stop after you have obtained six digits.*

(a) 0.9

(b) 0.4

(c) 0.6

(d) 0.83

(e) 0.25

(f) 0.44

(g) 0.482

(h) 0.667

(i) 0.315

Question 2.22 *Convert the following decimal numbers to hexadecimal numbers. Stop after you have obtained six digits.*

(a) 0.1

(b) 0.3

(c) 0.9

(d) 0.11

(e) 0.83

(f) 0.47

(g) 0.429

(h) 0.638

(i) 0.314

CHAPTER 3

Logic

Logic plays a fundamental role in many areas of computer science, including software engineering and design, expert systems, and artificial intelligence. A basic understanding of logic is necessary for many applications of computer science, as well as for many future courses.

3.1   PROPOSITIONS AND CONNECTIVES

In this chapter we introduce the basic laws of prepositional logic that are used in the design of computer systems. Prepositional logic is concerned with statements that can either be true or false. We learn how to analyse these statements and decide when they are true and when they are false.

Definition 3.1 *A **proposition** is a statement that is either true or false.*

Example 3.1

Some examples of statements that are propositions.

- Paris is the capital of Great Britain.
- Ten is greater than seven.
- Every even number is the sum of two prime numbers.

A proposition may be false, like the first proposition in the example, or it may be true, like the second proposition in the example, or you may not know if it is true or false, like the third proposition in the example. But think about the third proposition in the example, "Every even number is the sum of two prime numbers." It is clear that this statement can only be either true or false. If we know a proposition is true we say the truth-value of the proposition is true. If we know a proposition is false we say the truth-value of the propositions is false. Thus the **truth-value** of a proposition is either true or false.

Example 3.2

Some examples of statements that are not propositions.

- Where are you going?
- Close the door.
- This statement is false.

The first statement in this example is a question, it is neither true or false, it is a request for information. The second statement is a command or order, asking someone to do something. Again, it is neither true or false. The third statements is an example of a **paradox**. If we assume the statement "this statement is false" is true, then the statement itself says the statement must be false. If we assume "this statement is false" is false, then the statement itself says the statement must be true. While paradoxes can be interesting, we will not have anything more to say about them.

Example 3.3

Some ambiguous statements.

- Jon is tall.
- Chocolate is delicious.
- $x > 10$

Here are some statements that are ambiguous. That means it is not clear if they are propositions or not. How tall does one have to be to be considered tall? Some people think chocolate is delicious, some do not. And what is the variable x? Most books would say these statement are not propositions. Most of the time in computer science you will not encounter ambiguous statements like this so we will not worry about these kinds of statements.[1]

Propositions can be connected together using five words or phrases called **connectives**. There are five connectives that can be used to connect propositions:

1) **not (negation)**: ¬
2) **and**: ∧
3) **or**: ∨
4) **if-then (implies)**: →
5) **if-and-only-if (is-equivalent-to)**: ↔

Notice that each of these five connectives has a symbol associated with it. You will need to memorize these symbols. Please be aware that a few books use the symbol ∼ instead of ¬ for **not**, which is also sometimes called **negation**. The phrase "**if-then**" is also called "**implies**" and the phrase "**if-and-only-if**" is also called "**is-equivalent-to**." The five connectives can be used to connect propositions into **logical expressions**. Logical expressions are often just called **expressions**. They are also sometimes called **compound statements** or just **statements**. Just like in algebra where one can use a variable to represent a number, in logic we can use a variable to represent a proposition.

---

[1] These kinds of statements do play a role in advanced computer science, such as with artificial intelligence, fuzzy logic, probabilistic logic, and so on. But at our level we will not worry about them.

Example 3.4

Using the propositions $p = $ "Today is Monday." and $q = $ "It is raining." write the following logical expressions in words.

- $\neg p = $ Today is **not** Monday.

- $\neg q = $ It is **not** raining.

- $p \wedge q = $ Today is Monday **and** it is raining.

- $\neg p \wedge q = $ Today is **not** Monday **and** it is raining.

- $p \vee q = $ Today is Monday **or** it is raining.

- $p \vee \neg q = $ Today is Monday **or** it is **not** raining.

- $p \rightarrow q = $ **If** today is Monday **then** it is raining. = Today is Monday **implies** it is raining.

- $\neg p \rightarrow q = $ **If** today is **not** Monday **then** it is raining. = Today is **not** Monday **implies** it is raining.

- $p \leftrightarrow q = $ Today is Monday **if-and-only-if** it is raining. = Today is Monday **is-equivalent-to** it is raining.

- $\neg p \leftrightarrow q = $ Today is **not** Monday **if-and-only-if** it is raining. = Today is **not** Monday **is-equivalent-to** it is raining.

Here we have seen a few of the possible combinations of the connectives. Of course, the last example is a little silly. It is just meant to show you how to work with variables and the connectives. In computer science the propositions would be more serious.

You should notice when using connectives that **not** takes precedence. In other words, do the negation first. For example, if we have $p \wedge \neg q$ this really means $p \wedge (\neg q)$, we first find $\neg q$ and then we connect with $\wedge$. Also, in logic, parenthesis are used a lot like they are used in algebra, to help us determine the order in which we perform the operations. For example, in $p \wedge (q \vee r)$ we would do $q \vee r$ first before connecting p to it with $\wedge$.

Example 3.5

Let $p = $ "My program runs." and $q = $ "My program contains mistakes." Write the expression $(p \wedge \neg q) \vee q$ in words.

- My program runs **and** it contains **no** mistakes, **or** my program contains mistakes.
- Either my program runs **and** it contains **no** mistakes, **or** my program contains mistakes.

Both of these sentences are correct. We first do what is inside the parenthesis, $(p \wedge \neg q)$ and then connect it to q using $\vee$. Notice in the first sentence we use a comma to separate the part of the compound statement that is in parenthesis. In the second sentence we use the word "either" in addition to a comma to emphasize this separation. This is more a matter of writing clearly in English than logic.

3.2 CONNECTIVE TRUTH TABLES

Each of the five connectives has an associated **truth table**. A truth table is a table that helps us understand when logical expressions are true and when they are false. We begin with the simplest truth table, the truth table for the connective **not**:

p	$\neg p$
T	F
F	T

The variable p of course represents a proposition. (Do not be confused, it does not matter what letter we use as the variable. We could have used q or r or x or something else.) The first row of the table shows that if p is true then $\neg p$ is false. The second row shows that if p is false then $\neg p$ is true. This should be obvious to you. If a proposition is true then the negation of that proposition must be false. Similarly, if a proposition is false then the negation of that proposition must be true.

> ### Example 3.6
>
> Let $p =$ "$2 + 2 = 4$." Clearly p is true. The negation of p is $\neg p =$ "$2 + 2 \neq 4$," which is clearly false. If we let $q =$ "$3 + 2 = 6$" then clearly q is false. The negation of q is $\neg q =$ "$3 + 2 \neq 6$," which is clearly true.

The next truth table is for the connective **and**:

p	q	$p \wedge q$
T	T	T
T	F	F
F	T	F
F	F	F

Since the connective **and** actually connects two propositions, the truth table for **and** has two propositions in it, p and q. The proposition p could be either true or false. Similarly, the proposition q could be either true or false. A truth table must have enough rows to include every possible combination of true or false for the variables in the table. Therefore the number of rows a truth table has is equal to 2^n where n is the number of proposition variables we have. Here we have two proposition variables, p and q, so the table must have $2^2 = 4$ rows in it. According to the truth table, the only way to have $p \wedge q$ be true is if both p and q are true. If we think about this in terms of English sentences it makes sense.

> ### Example 3.7
>
> Let
>
> $p =$ Paris is in France. (**True**)   and $q =$ London is in England. (**True**)
>
> The whole statement
>
> $p \wedge q =$ Paris is in France **and** London is in England.
>
> is true. But suppose instead we had
>
> $p =$ Paris is in France. (**True**)   and $q =$ London is in China. (**False**)

then the whole statement

$$p \wedge q = \text{Paris is in France } \textbf{and} \text{ London is in China.}$$

is false since one part of the statement is false. Similarly, if we had

$$p = \text{Paris is in Japan. } (\textbf{False}) \quad \text{and} \quad q = \text{London is in England. } (\textbf{True})$$

then the whole statement

$$p \wedge q = \text{Paris is in Japan } \textbf{and} \text{ London is in England.}$$

is false since one part of the statements is false. Finally, if we had

$$p = \text{Paris is in Japan. } (\textbf{False}) \quad \text{and} \quad q = \text{London is in China. } (\textbf{False})$$

then the whole statement

$$p \wedge q = \text{Paris is in Japan } \textbf{and} \text{ London is in China.}$$

is false since both parts of the statements are false.

Next we will consider the connective **or**. However, with this word we have to be careful. In English the word or can be used in two distinct ways. Consider the following two sentences:

- You may have either tea or coffee with breakfast. (**exclusive-or**)
- The movie has a discount for students or children. (**inclusive-or**)

If you are at a restaurant and the menu says you may have either tea or coffee with your breakfast, then you are expected to either choose tea or choose coffee. You cannot choose both. The **or** in this situation is called an **exclusive-or**. You can chose one or the other but not both. If a movie has a discount for students or children then what about children who are students? The discount is available to people who are students, it is available to people who are children, and it is available to people who are both students and children. In this situation the **or** is called an **inclusive-or**. The **connective** ∨ **refers to the inclusive-or**. The expression $q \vee p$ is true if p is true, if q is true, or if both p and q are true. This helps us understand the truth table for **or**:

p	q	$p \vee q$
T	T	T
T	F	T
F	T	T
F	F	F

Example 3.8

Let

$$p = \text{Paris is in France. } (\textbf{True}) \quad \text{and} \quad q = \text{London is in England. } (\textbf{True})$$

The whole statement

$$p \vee q = \text{Paris is in France } \textbf{or} \text{ London is in England.}$$

is clearly true since both parts of the statement is true. But suppose we have

$$p = \text{Paris is in France. } (\textbf{True}) \quad \text{and} \quad q = \text{London is in China. } (\textbf{False})$$

The whole statement

$$p \lor q = \text{Paris is in France } \textbf{or} \text{ London is in China.}$$

is still true since one of the parts of the statement is true. Similarly, if we have

$$p = \text{Paris is in Japan. } (\textbf{False}) \quad \text{and} \quad q = \text{London is in England. } (\textbf{True})$$

The whole statement

$$p \lor q = \text{Paris is in Japan } \textbf{or} \text{ London is in England.}$$

is true since again one of the parts of the statement is true. Finally, suppose we had

$$p = \text{Paris is in Japan. } (\textbf{False}) \quad \text{and} \quad q = \text{London is in China. } (\textbf{False})$$

then the whole statement

$$p \lor q = \text{Paris is in Japan } \textbf{or} \text{ London is in China.}$$

is false since both parts of the statement is false.

Example 3.9

Though we will not use it in this chapter, the **exclusive-or** is important as well in both logic and Boolean algebra. We simply use **xor** to represent the **exclusive-or**. The expression p **xor** q is true when only one of the propositions p or q is true. That means that if both p and q are true, then p **xor** q is false. Here is the truth table for the **exclusive-or**.

p	q	p **xor** q
T	T	F
T	F	T
F	T	T
F	F	F

In this case the expression

$$p \textbf{ xor } q = \text{Paris is in France } \textbf{or} \text{ London is in England.}$$

would be considered false since both of the propositions are true.

The next two connectives, **if-then (implies)** and **if-and-only-if (is-equivalent-to)**, can seem a little strange if we rely too heavily on language for our understanding. We can come up with some very strange examples that do not make much sense. This is because logic in the real world is usually applied to mathematical propositions, not silly propositions like "Paris is in France" or "Today is Monday." Logic, when applied to mathematical propositions, works perfectly.

Next we will consider the truth table for **if-and-only-if**. Another way of saying **if-and-only-if** is to say **is-equivalent-to**. In English two things are considered to be equivalent if they are the same in some way. In logic what is important is the truth-value of propositions or expressions. Therefore, in logic two statements are considered equivalent if they have the same truth-value. In other words, two statements are considered equivalent if they are either both true or both false. Thus the truth table for **if-and-only-if** is given by:

p	q	$p \leftrightarrow q$
T	T	T
T	F	F
F	T	F
F	F	T

Example 3.10

Let us consider the two true propositions

$$p = \text{Paris is in France. (\textbf{True})} \quad \text{and} \quad q = \text{London is in England. (\textbf{True})}$$

The whole statement

$$p \leftrightarrow q = \text{Paris is in France \textbf{if-and-only-if} London is in England.}$$
$$= \text{Paris is in France \textbf{is-equivalent-to} London is in England.}$$

is true. But as a sentence in English it does not make a whole lot of sense. Why should we expect the locations of Paris and London to be related to each other? Now suppose we have

$$p = \text{Paris is in France. (\textbf{True})} \quad \text{and} \quad q = \text{London is in China. (\textbf{False})}$$

The whole statement

$$p \leftrightarrow q = \text{Paris is in France \textbf{if-and-only-if} London is in China.}$$
$$= \text{Paris is in France \textbf{is-equivalent-to} London is in China.}$$

is now false since one part of the statement is true and the other part is false. But again, the two parts of the statement seem unrelated to each other. Similarly, if we had

$$p = \text{Paris is in Japan. (\textbf{False})} \quad \text{and} \quad q = \text{London is in England. (\textbf{True})}$$

The whole statement

$$p \leftrightarrow q = \text{Paris is in Japan \textbf{if-and-only-if} London is in England.}$$
$$= \text{Paris is in Japan \textbf{is-equivalent-to} London is in England.}$$

is also false since one part of the statement is true and the other part is false. But again, the two parts of the statement seem unrelated to each other. Finally, suppose we had

$$p = \text{Paris is in Japan. (\textbf{False})} \quad \text{and} \quad q = \text{London is in China. (\textbf{False})}$$

then the whole statement

$$p \leftrightarrow q = \text{Paris is in Japan \textbf{if-and-only-if} London is in China.}$$
$$= \text{Paris is in Japan \textbf{is-equivalent-to} London is in China.}$$

is true according to our table since both parts of the statement are false. But again, as an English sentence it makes very little sense. Just remember, when using logic in real life, in problems that come from computer science or mathematics, logic works perfectly.

We can convince ourselves that the **if-and-only-if** truth table works since we can use the argument that true statements are equivalent in some sense and false statements are equivalent in some sense. In other words, it is the truth-value of the propositions that is important. We will also use this idea to try to understand the **if-then** truth table. In the statement "$p \rightarrow q$" the first part of the statement, p, is called the **premise** and the second part of the statement, q, is called the **conclusion**. Here is one way to think about it:

- A true premise can only imply a true conclusion. (So if p is true and q is true then $p \rightarrow q$ is also true.)
- A true premise can never imply a false conclusion. (So if p is true and q is false then $p \rightarrow q$ must be false.)
- A false premise, because it is false, can imply anything. (So if p is false and q is true then $p \rightarrow q$ is true. But also, if p is false and q is false then $p \rightarrow q$ is also true.)

This gives the following truth table for the **if-then** connective:

p	q	$p \rightarrow q$
T	T	T
T	F	F
F	T	T
F	F	T

Example 3.11

Let

$$p = \text{Paris is in France. (\textbf{True})} \quad \text{and} \quad q = \text{London is in England. (\textbf{True})}$$

then the whole statement

$$p \rightarrow q = \textbf{If} \text{ Paris is in France \textbf{then} London is in England.}$$
$$= \text{Paris is in France \textbf{implies} London is in England.}$$

is true. But again as a sentence in English this does not make a whole lot of sense. Why should we expect the locations of Paris and London to be related to each other? Now suppose we have

$$p = \text{Paris is in France. (\textbf{True})} \quad \text{and} \quad q = \text{London is in China. (\textbf{False})}$$

The whole statement

$$p \to q = \textbf{If} \text{ Paris is in France } \textbf{then} \text{ London is in China.}$$
$$= \text{Paris is in France } \textbf{implies} \text{ London is in China.}$$

is now false since the premise, or first part of the statement, is true and the conclusion, or second part of the statement, is false. But again, the two parts of the statement seem unrelated to each other. Now suppose we had

$$p = \text{Paris is in Japan. } (\textbf{False}) \quad \text{and} \quad q = \text{London is in England. } (\textbf{True})$$

then the whole statement

$$p \to q = \textbf{If} \text{ Paris is in Japan } \textbf{then} \text{ London is in England.}$$
$$= \text{Paris is in Japan } \textbf{implies} \text{ London is in England.}$$

is true according to our table since the premise is false and the conclusion is true. The fact that the whole statement is considered to be true just seems silly with this example. Now suppose

$$p = \text{Paris is in Japan. } (\textbf{False}) \quad \text{and} \quad q = \text{London is in China. } (\textbf{False})$$

then the whole statement

$$p \to q = \textbf{If} \text{ Paris is in Japan } \textbf{then} \text{ London is in China.}$$
$$= \text{Paris is in Japan } \textbf{implies} \text{ London is in China.}$$

is true according to our table since the premise is false and the conclusion is false. Again, the fact that the whole statement is considered to be true just seems silly with this example. But when using logic in problems that come from computer science or mathematics, everything works perfectly.

3.3   TRUTH VALUE OF COMPOUND STATEMENTS

When we are given a compound statement we usually want to know if or when the compound statement is true or false. This can be done in one of two ways. If we already know the truth-value of each proposition then we can find out directly. If we are not given the truth-value of each proposition we need to use a truth table.

When the truth-values of all the individual propositions in a compound statement are known, we can find out the truth-value of the compound statement directly. We substitute the truth-values of the propositions into the compound statement and use the connective truth-tables. Remember that parenthesis in logic are used just like parenthesis in algebra, they tell us in what order to evaluate things.

Example 3.12

If p is true and q is false find the truth value of $(p \wedge \neg q) \vee q$.

$$(T \wedge \neg F) \vee F \qquad \text{substitute truth-values}$$
$$\equiv (T \wedge T) \vee F \qquad \text{use truth-table of } \neg$$
$$\equiv T \vee F \qquad \text{use truth-table of } \wedge$$
$$\equiv T \qquad \text{use truth-table of } \vee$$

The symbol $\equiv$ is read "is equivalent to." This is different than the **if-and-only-if** connective which is also known as **is-equivalent-to**. Here the symbol $\equiv$ operates a lot like the equal sign in algebra. In algebra you would write $2(x + y) = 2x + 2y$ to say that both $2(x + y)$ and $2x + 2y$ are the same expression just written in different ways. In logic we use $\equiv$ to say both sides have the same truth-value.

Example 3.13

If q is true find the truth value of $\neg(\neg q \wedge \neg q)$.

$$\neg(\neg T \wedge \neg T)$$
$$\equiv \neg(F \wedge F)$$
$$\equiv \neg F$$
$$\equiv T$$

Example 3.14

If p and q are both false find the truth value of $\neg(p \wedge q) \rightarrow (\neg q \vee \neg q)$.

$$\neg(F \wedge F) \rightarrow (\neg F \vee \neg F)$$
$$\equiv \neg(F \wedge F) \rightarrow (T \vee T)$$
$$\equiv \neg F \rightarrow T$$
$$\equiv T \rightarrow T$$
$$\equiv T$$

Often you are asked simply to find the **truth table** of a compound statement. This means you are to find the truth-value of the compound statement for every possible proposition truth-value. You need to make and fill-out a truth-table. The number of rows your truth table needs is 2^n where n is the number of individual propositions your compound statement has. The truth table for $\neg(p \wedge q)$ would need $2^2 = 4$ rows since it has two propositions (p and q); the truth table for $(\neg p \wedge q) \vee r$ would need $2^3 = 8$ rows since it has three propositions (p, q, and r); and the truth table for $\neg(p \vee q) \rightarrow (\neg r \wedge s)$ would need $2^4 = 16$ rows since it has four propositions (p, q, r, and s). To figure out the columns a truth table needs it may be helpful to make an **expression tree**. (Trees will be studied later in chapter 10.)

Example 3.15

Make an expression tree for $\neg(p \wedge q)$ and use that to construct a truth table.

The construction of this expression tree should be clear, in order to find $\neg(p \wedge q)$ we must first find $(p \wedge q)$ and then take the negation of that. We start at the bottom of the tree where the variables p and q each appear in a circle called a **vertex**.[a] These variables are then combined with $\wedge$ which is in the vertex above. The "combining" is shown by connecting the variables p and q to $\wedge$ by lines called **edges**. The result is $p \wedge q$. The result of this "combining" is then negated, which happens at the top vertex. This negation is shown by connecting $\wedge$ to $\neg$ with an edge. Thus we get $\neg(p \wedge q)$. The truth table requires a column for each vertex in the expression tree.

p	q	$p \wedge q$	$\neg(p \wedge q)$
T	T		
T	F		
F	T		
F	F		

In the first two columns of the truth table we have filled in every possible combination of truth values for p and q. The rest of the truth table needs to be filled in.

[a]The plural of vertex is either vertices or vertexes.

Example 3.16

Make an expression tree for $(p \wedge \neg q) \vee q$ and use that to construct a truth table.

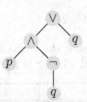

In the expression tree the variables are the vertices at the bottom. The first thing we do is find $\neg q$. We then $\wedge$ together p and $\neg q$ to get $p \wedge \neg q$. Finally, we have to $\vee$ this together with q to get $(p \wedge \neg q) \vee q$. We make a truth table using the vertices of the expression tree; each vertex results in a column in the truth table. Notice, in the truth table we do not need to put the variable q in twice, once is enough.

p	q	$\neg q$	$p \wedge \neg q$	$(p \wedge \neg q) \vee q$
T	T			
T	F			
F	T			
F	F			

Example 3.17

Make an expression tree for $\neg(\neg p \wedge q) \leftrightarrow (\neg q \vee r)$ and use that to construct a truth table.

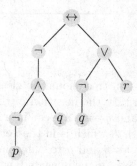

Study $\neg(\neg p \wedge q) \leftrightarrow (\neg q \vee r)$ and the expression tree until the construction of the tree is clear. We make the columns in the truth table using the vertices of the expression tree. We do not need to put the variable q in twice.

p	q	r	$\neg p$	$\neg p \wedge q$	$\neg(\neg p \wedge q)$	$\neg q$	$\neg q \vee r$	$\neg(\neg p \wedge q) \leftrightarrow (\neg q \vee r)$
T	T	T						
T	T	F						
T	F	T						
T	F	F						
F	T	T						
F	T	F						
F	F	T						
F	F	F						

Example 3.18

Fill out the truth table for $(p \wedge \neg q) \vee q$.

First we fill in the third column using the second column and the truth table for $\neg$. When q is true then $\neg q$ is false, and when q is false then $\neg q$ is true.

p	q	$\neg q$	$p \wedge \neg q$	$(p \wedge \neg q) \vee q$
T	T	F		
T	F	T		
F	T	F		
F	F	T		

Next we fill in the fourth column using the first and third columns and the truth table for $\land$.

p	q	$\neg q$	$p \land \neg q$	$(p \land \neg q) \lor q$
T	T	F	F	
T	F	T	T	
F	T	F	F	
F	F	T	F	

Finally we fill in the fifth column using the second and fourth columns and the truth table for $\lor$.

p	q	$\neg q$	$p \land \neg q$	$(p \land \neg q) \lor q$
T	T	F	F	T
T	F	T	T	T
F	T	F	F	T
F	F	T	F	F

With the truth table we can see exactly when the compound statement $(p \land \neg q) \lor q$ is true and when it is false. It is false when both p and q are false and true otherwise. Notice this is exactly the same as the truth table for $p \lor q$.

3.4   TAUTOLOGIES AND CONTRADICTIONS

A compound statement, or expression, is called a **tautology** if it is always true, regardless of the proposition truth-values. An expression is called a **contradiction** if it is always false, regardless of the proposition truth-values.

Example 3.19

Show that $\neg(p \land q) \to (\neg p \lor \neg q)$ it a tautology. We first construct the expression tree.

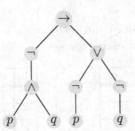

We use the expression tree to construct the truth table which we then fill out.

p	q	$p \land q$	$\neg(p \land q)$	$\neg p$	$\neg q$	$\neg p \lor \neg q$	$\neg(p \land q) \to (\neg p \lor \neg q)$
T	T	T	F	F	F	F	T
T	F	F	T	F	T	T	T
F	T	F	T	T	F	T	T
F	F	F	T	T	T	T	T

We can see that the expression $\neg(p \land q) \to (\neg p \lor \neg q)$ is true regardless of the truth-values of the propositions. Thus $\neg(p \land q) \to (\neg p \lor \neg q)$ is called a tautology. We sometimes say that $\neg(p \land q) \to (\neg p \lor \neg q)$ is **tautologically** true.

Example 3.20

Show that $(p \wedge q) \wedge \neg p$ is a contradiction. We first make the expression tree. If you feel comfortable making the truth table without making the expression tree then there is no need to make the tree.

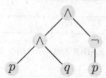

We then use the expression tree to construct the truth table which we then fill out.

p	q	$p \wedge q$	$\neg p$	$(p \wedge q) \wedge \neg p$
T	T	T	F	F
T	F	F	F	F
F	T	F	T	F
F	F	F	T	F

We can see that the expression $(p \wedge q) \wedge \neg p$ is false regardless of the truth-value of the propositions. Thus $(p \wedge q) \wedge \neg p$ is called a contradiction. A contradiction is never true.

3.5 LOGICAL EQUIVALENCE AND THE LAWS OF LOGIC

Two expressions that contain the same variables are called **logically equivalent** if they have the same truth-values for every combination of variable truth-values. The symbol used for this is $\equiv$. In logic, the symbol $\equiv$ operates a lot like the symbol $=$ in algebra. The symbol $\equiv$ is read "is equivalent to," but do not confuse this with the connective **is-equivalent-to**.

Example 3.21

Show that the expressions $\neg(p \wedge q)$ and $\neg p \vee \neg q$ are logically equivalent. This requires constructing the truth table for each expression.

p	q	$p \wedge q$	$\neg(p \wedge q)$
T	T	T	F
T	F	F	T
F	T	F	T
F	F	F	T

and

p	q	$\neg p$	$\neg q$	$\neg p \vee \neg q$
T	T	F	F	F
T	F	F	T	T
F	T	T	F	T
F	F	T	T	T

Notice how the two expressions have the same truth-values for every combination of variable truth-value. When both p and q are true then both expressions are false. When p is true and q is false then both expressions are true. When p is false and q is true then both expressions are true. When both p and q are false then both expressions are false. Thus the expression $\neg(p \wedge q)$ is logically equivalent to the expression $\neg p \vee \neg q$. This is written as

$$\neg(p \wedge q) \equiv \neg p \vee \neg q.$$

This equivalence is the first of **DeMorgan's laws**.

Example 3.22

Show that the expressions $\neg(p \lor q)$ and $\neg p \land \neg q$ are logically equivalent. We do this by constructing truth tables for the two expressions.

p	q	$p \lor q$	$\neg(p \lor q)$
T	T	T	F
T	F	T	F
F	T	T	F
F	F	F	T

and

p	q	$\neg p$	$\neg q$	$\neg p \land \neg q$
T	T	F	F	F
T	F	F	T	F
F	T	T	F	F
F	F	T	T	T

Again, the two expressions have the same truth-values for every combination of variable truth-value. When both p and q are true then both expressions are false. When p is true and q is false then both expressions are false. When p is false and q is true then both expressions are false. When both p and q are false then both expressions are true. Thus the expression $\neg(p \lor q)$ is logically equivalent to the expression $\neg p \land \neg q$,

$$\neg(p \lor q) \equiv \neg p \land \neg q.$$

This equivalence is the second of **DeMorgan's laws**.

Example 3.23

Show that the expressions $p \to q$ and $\neg q \to \neg p$ are equivalent. The expression $\neg q \to \neg p$ is called the **contrapositive** of $p \to q$. We make the truth tables. The first is simply the truth-table for the if-then connective.

p	q	$p \to q$
T	T	T
T	F	F
F	T	T
F	F	T

and

p	q	$\neg p$	$\neg q$	$\neg q \to \neg p$
T	T	F	F	T
T	F	F	T	F
F	T	T	F	T
F	F	T	T	T

Thus we see that these two expressions are equivalent. We write

$$p \to q \equiv \neg q \to \neg p.$$

This may surprise you, but this equivalence is actually one of the most important relationships in mathematics and computer science. As we said, the expression $\neg q \to \neg p$ is called the **contrapositive** of $p \to q$. It is a fundamental way that mathematicians and computer scientists prove things. Many times you want to show a statement $p \to q$ is true but showing this statement is true is difficult while it is easy to show the contrapositive $\neg q \to \neg p$ is true. But since these two statements are equivalent then showing the contrapositive of $p \to q$ is true is the same thing as showing that $p \to q$ is true.

There are two special laws that are almost like definitions instead of laws. They allow us to write $\to$ and $\leftrightarrow$ in terms of $\neg$, $\land$, and $\lor$. These can be shown to be true in exactly the same way as in the last example.

Name	Laws
Implication law	$p \to q \equiv \neg p \lor q$
Equivalence law	$p \leftrightarrow q \equiv (p \to q) \land (q \to p) \equiv (\neg p \lor q) \land (\neg q \lor p)$

There is also a standard list of equivalence relations that are often called the **laws of logic**. Each of these laws of logic can be shown to be true just as in the examples above. We can construct the truth tables for the expression on each side of the $\equiv$ sign and compare them. It is a good idea for you to do that for these laws and convince yourself that they are true. We have already done this for both of DeMorgan's laws.

Name	Laws of Logic	
Commutative laws	(a) $p \wedge q \equiv q \wedge p$	(b) $p \vee q \equiv q \vee p$
Associative laws	(a) $(p \wedge q) \wedge r \equiv p \wedge (q \wedge r)$	(b) $(p \vee q) \vee r \equiv p \vee (q \vee r)$
Distributive laws	(a) $p \wedge (q \vee r) \equiv (p \wedge q) \vee (p \wedge r)$	(b) $p \vee (q \wedge r) \equiv (p \vee q) \wedge (p \vee r)$
Identity laws	(a) $p \wedge T \equiv p$	(b) $p \vee F \equiv p$
Inverse laws	(a) $p \wedge \neg p \equiv F$	(b) $p \vee \neg p \equiv T$
Double negation law	$\neg \neg p \equiv p$	
Idempotent laws	(a) $p \wedge p \equiv p$	(b) $p \vee p \equiv p$
DeMorgan's laws	(a) $\neg (p \wedge q) \equiv \neg p \vee \neg q$	(b) $\neg (p \vee q) \equiv \neg p \wedge \neg q$
Annihilation laws	(a) $p \wedge F \equiv F$	(b) $p \vee T \equiv T$
Absorption laws	(a) $p \wedge (p \vee q) \equiv p$	(b) $p \vee (p \wedge q) \equiv p$
Negation laws	(a) $\neg T = F$	(b) $\neg F = T$

The reason the laws of logic are so important is that we can often use them to simplify complicated compound expressions. Doing this feels a lot like simplifying expressions in algebra, but you have to be careful. The laws of logic are a bit different than the laws of algebra. You want to make sure that you can justify each step with a law of logic.

Example 3.24

Simplify the expression $(p \vee \neg q) \wedge (p \vee q)$ using the laws of logic. Justify each step.

$$(p \vee \neg q) \wedge (p \vee q)$$
$$\equiv p \vee (\neg q \wedge q) \qquad\qquad \text{Distributive law}$$
$$\equiv p \vee (q \wedge \neg q) \qquad\qquad \text{Commutative law}$$
$$\equiv p \vee F \qquad\qquad\qquad \text{Inverse law}$$
$$\equiv p \qquad\qquad\qquad\quad \text{Identity law}$$

Example 3.25

Simplify the expression $\neg[p \rightarrow \neg(p \wedge q)]$ using the laws of logic. Justify each step.

$$\neg[p \rightarrow \neg(p \wedge q)]$$
$$\equiv \neg[\neg p \vee \neg(p \wedge q)] \qquad\qquad \text{Implication law}$$
$$\equiv \neg\neg p \wedge \neg\neg(p \wedge q) \qquad\qquad \text{DeMorgan's law}$$
$$\equiv p \wedge (p \wedge q) \qquad\qquad\qquad \text{Double negation law}$$
$$\equiv (p \wedge p) \wedge q \qquad\qquad\qquad \text{Associative law}$$
$$\equiv p \wedge q \qquad\qquad\qquad\qquad \text{Idempotent law}$$

Example 3.26

Simplify the expression $\neg\left[p \vee (q \wedge \neg p)\right]$ using the laws of logic. Justify each step.

$$\neg\left[p \vee (q \wedge \neg p)\right]$$

$\equiv \neg p \wedge \neg(q \wedge \neg p)$	DeMorgan's law
$\equiv \neg p \wedge (\neg q \vee \neg\neg p)$	DeMorgan's law
$\equiv \neg p \wedge (\neg q \vee p)$	Double negation law
$\equiv \neg p \wedge (p \vee \neg q)$	Commutative law
$\equiv (\neg p \wedge p) \vee (\neg p \wedge \neg q)$	Distributive law
$\equiv F \vee (\neg p \wedge \neg q)$	Inverse law
$\equiv \neg p \wedge \neg q$	Identity law

3.6 PROBLEMS

Question 3.1 *Are the following statements propositions?*
- (a) *Cats eat a lot.*
- (b) *How are you?*
- (c) *Paris is in China*
- (d) $17 = 16$
- (e) $7 > 13$
- (f) $12 + 15 = 27$
- (g) *Go fly a kite.*
- (h) *Close the door!*
- (i) *Roses are flowers.*

Question 3.2 *Let $p =$ "Jon plays baseball" and $q =$ "Ron plays football." Write the following logical statements in words.*
- (a) $p \vee q$
- (b) $q \wedge p$
- (c) $\neg p \wedge q$
- (d) $p \wedge \neg q$
- (e) $\neg q \vee \neg p$
- (f) $\neg(q \vee p)$

Question 3.3 *Suppose p is true and q is false. Find the truth value of the following compound statements.*
- (a) $p \wedge q$
- (b) $q \wedge p$
- (c) $\neg q \vee p$
- (d) $\neg p \vee q$
- (e) $p \wedge \neg q$
- (f) $\neg q \vee \neg p$
- (g) $p \to q$
- (h) $\neg q \to p$
- (i) $\neg q \to \neg p$
- (j) $p \leftrightarrow q$
- (k) $\neg q \leftrightarrow p$
- (l) $\neg p \leftrightarrow \neg q$

Question 3.4 *Construct the truth table for the following compound statements.*
- (a) $p \wedge q$
- (b) $q \wedge p$
- (c) $\neg q \vee p$
- (d) $\neg p \vee q$
- (e) $p \wedge \neg q$
- (f) $\neg q \vee \neg p$
- (g) $p \to q$
- (h) $\neg q \to p$
- (i) $\neg q \to \neg p$
- (j) $p \leftrightarrow q$
- (k) $\neg q \leftrightarrow p$
- (l) $\neg p \leftrightarrow \neg q$

Question 3.5 *Show the Implication law $p \to q \equiv \neg p \vee q$ by constructing a truth table for each side of the implication and showing these two truth tables are equivalent.*

Question 3.6 *Show the Equivalence law $p \leftrightarrow q \equiv (p \to q) \wedge (q \to p)$ by constructing a truth table for each side of the implication and showing these two truth tables are equivalent. Then use the Implication law to rewrite $p \leftrightarrow q \equiv (p \to q) \wedge (q \to p)$ in terms of $\wedge$, $\vee$, and $\neg$ only.*

Question 3.7 *Construct the truth table for the following compound statements.*

(a) $(p \vee \neg r) \to q$

(b) $(q \wedge p) \to \neg p$

(c) $\neg p \to (\neg p \vee q)$

(d) $\neg(p \wedge \neg q) \vee r$

(e) $\neg[(\neg p \wedge q) \vee r]$

(f) $(p \wedge \neg q) \wedge (r \vee q)$

(g) $(p \vee q) \leftrightarrow [p \vee (q \wedge r)]$

(h) $(q \wedge \neg r) \leftrightarrow \neg(p \vee r)$

(i) $(p \leftrightarrow q) \wedge (r \leftrightarrow q)$

Question 3.8 *(Easy) Simplify the following Logical expressions with the laws of logic.*

(a) $p \vee p$

(b) $p \vee \neg p$

(c) $\neg(p \vee \neg p)$

(d) $p \vee (p \wedge q)$

(e) $p \vee (\neg p \wedge q)$

(f) $p \wedge (\neg p \vee q)$

(g) $p \wedge p$

(h) $p \wedge (p \vee q)$

(i) $p \wedge (p \vee q \vee r)$

(j) $(p \wedge q) \vee (p \wedge \neg q)$

(k) $\neg(\neg p \vee \neg p)$

(l) $(p \wedge q) \vee (\neg p \wedge q)$

Question 3.9 *(Moderate) Simplify the following Logical expressions with the laws of logic.*

(a) $(\neg p \vee \neg q) \wedge (\neg p \vee q)$

(b) $q \vee (q \wedge \neg q)$

(c) $\neg p \vee (q \wedge \neg p)$

(d) $(p \vee \neg q) \wedge (p \vee q)$

(e) $r \vee [r \vee (r \wedge p)]$

(f) $p \wedge [p \vee (p \wedge q)]$

(g) $r \vee (r \wedge \neg p \wedge q \wedge s)$

(h) $\neg r \wedge \neg(r \wedge p \wedge q \wedge s)$

Question 3.10 *(Difficult) Simplify the following Logical expressions with the laws of logic.*

(a) $q \wedge [p \vee (\neg p \wedge q)]$

(b) $\neg[(p \wedge \neg q) \vee \neg p]$

(c) $(\neg p \wedge q) \vee [p \wedge (p \vee q)]$

(d) $\neg[p \vee (p \wedge q)] \wedge q$

(e) $(p \wedge \neg r) \vee (\neg p \wedge q) \vee \neg(q \wedge r)$

(f) $\neg(p \wedge q) \vee (q \wedge r)$

(g) $[p \vee (q \wedge r)] \wedge (\neg p \vee r)$

(h) $(p \wedge \neg r) \vee (\neg p \wedge q) \vee \neg(p \wedge r)$

(i) $(p \wedge r) \vee (\neg p \wedge q) \vee (r \wedge q)$

(j) $\neg p \vee \neg q \vee [p \wedge q \wedge \neg r]$

(k) $(p \vee r) \wedge (\neg p \vee q) \wedge (r \vee q)$

(l) $\neg(p \wedge q) \wedge (\neg p \vee q) \wedge (\neg q \vee q)$

Question 3.11 *Which of the following statements are tautologies?*

(a) $(p \wedge \neg q) \to \neg q$

(b) $(\neg p \wedge q) \to \neg q$

(c) $(\neg p \wedge q) \to \neg p$

(d) $(\neg p \vee q) \to \neg p$

Question 3.12 *Which of the following statements are contradictions?*

(a) $(p \vee q) \wedge (\neg p \wedge \neg q)$

(b) $(p \vee q) \wedge (p \wedge \neg q)$

(c) $(\neg p \vee \neg q) \wedge (\neg p \wedge q)$

(d) $(p \vee \neg q) \wedge (\neg p \wedge q)$

CHAPTER 4

Set Theory

An understanding of sets and set notation is necessary for you in the future. The language of set theory is used throughout computer science. Relations between sets are also very important in some areas of computer science, especially when studying databases. We will not cover databases in this course, but we will cover the basic definitions and ideas that you will need to know later.

4.1   SET NOTATION

A **set** is a **well-defined** collection of objects. The objects in a set are usually called the **elements** of the set, though sometimes they are also called the **members** of the set. The phrase *well-defined* just means that it is clear if an object is in the set or not. Sets are usually indicated by curly-brackets { }.

Example 4.1

Some examples of sets.

- {apple, orange, banana} is a set that has three elements, the words apple, orange, and banana.

- $\{1, 2, 3, 4, 5\}$ is a set that has five elements, the numbers 1, 2, 3, 4, and 5.

- $\{1, 2, 3, \ldots, 50\}$ is a set that has fifty elements, the positive integers 1 through 50.

- $\{2, 4, 6, 8, 10, 12, \ldots\}$ is a set that contains all the positive even integers. This set has an infinite number of elements since there are an infinite number of positive even integers.

- $\{a, b, c\}$ is a set that contains the letters a, b, and c.

- $\{a, b, c, \ldots, z\}$ is a set that contains the lowercase letters of the alphabet.

If x is an element of the set S then we write $x \in S$. They symbol $\in$ is read "is an element of." If x is not an element of set S then we would write $x \notin S$. The symbol $\notin$ is read "is not an element of."

Example 4.2

Elements of sets.

- apple $\in \{\text{apple}, \text{orange}, \text{banana}\}$ but pear $\notin \{\text{apple}, \text{orange}, \text{banana}\}$.
- $4 \in \{1, 2, 3, 4, 5\}$ but $6 \notin \{1, 2, 3, 4, 5\}$.
- $13 \in \{1, 2, 3, \ldots, 50\}$ but $157 \notin \{1, 2, 3, \ldots, 50\}$.
- $8342 \in \{2, 4, 6, 8, 10, 12, \ldots\}$ but $6329 \notin \{2, 4, 6, 8, 10, 12, \ldots\}$.

Now we will introduce some of the most important sets in mathematics and computer science.

$$\begin{aligned}
\mathbb{N} = {}& \text{the set of } \textbf{natural numbers} \\
= {}& \text{the set of positive integers} \\
= {}& \{1, 2, 3, 4, \ldots\}, \\
\mathbb{Z} = {}& \text{the set of } \textbf{integer numbers} \\
= {}& \{\ldots, -4, -3, -2, -1, 0, 1, 2, 3, 4, \ldots\}, \\
\mathbb{Q} = {}& \text{the set of } \textbf{rational numbers} \\
= {}& \left\{ x \mid x = \frac{m}{n} \text{ where } m, n \in \mathbb{Z} \text{ and } n \neq 0 \right\}, \\
\mathbb{R} = {}& \text{the set of } \textbf{real numbers}.
\end{aligned}$$

Sometimes the natural numbers[1] are also called the **whole numbers** or **counting numbers**. Also, the integer numbers are often simply called the **integers**, the rational numbers are often simply called the **rationals**, and the real numbers are often simply called the **reals**. It is difficult to give a precise definition of the real numbers, but you should be familiar with them as the real number line. That is a good way to think about the real numbers.

Also, notice how we defined the rational numbers $\mathbb{Q}$. Sets can be written in one of two different ways. They can be written in an **enumerated form** or in a **predicate form**. Most of the sets above were written in an enumerated form. To enumerate simply means to list, and in most of the above examples you can see that we simply listed the elements in the set inside the curly-brackets. Writing a set in predicate form means we use an expression to help us define the set. This is how we defined $\mathbb{Q}$ above. A **predicate** is a statement involving one or more variables such that when we assign values to the variables the statement becomes a proposition which is either true for false. We write $P(x)$ to mean a predicate of x and read it as "pee-of-ex."

Sets in predicate form are written either as

$$\{x \mid P(x)\},$$

which is read as "the set of x such that $P(x)$ is true," or

$$\{x \in S \mid P(x)\},$$

which is read as "the set of x in S such that $P(x)$ is true."

---

[1] Often the set of natural numbers also includes zero. From a computer science perspective this makes sense in terms of indexing arrays.

The curly-brackets tells us we have a set. The x tells us the name of the variable, and if we have written $x \in S$ then that tells us that x comes from set S. The $|$ is read as "such that." Some books use a : instead of a $|$ for "such that." The $P(x)$ tells us the condition that must be satisfied for a value of x to be in the set. This means that $P(x)$ must be true for a value of x in order for x to be in the set. First we look at examples of predicates before looking at examples of sets written in predicate form.

Example 4.3

Suppose that $x \in \mathbb{Z}$. The statement

$$5 \leq x \leq 10$$

is a predicate of x. If we assign the value 7 to the variable x the statement becomes the proposition $5 \leq 7 \leq 10$, which is true. If we assign the value -2 to the variable x the statement becomes the proposition $5 \leq -2 \leq 10$, which is false.

Example 4.4

Suppose that $x \in \mathbb{R}$. The statement

$$x = \frac{m}{n} \text{ where } m, n \in \mathbb{Z} \text{ and } n \neq 0$$

is a predicate in x. If we assign the value 3.74 to the variable x we have the statement

$$3.74 = \frac{m}{n} \text{ where } m, n \in \mathbb{Z} \text{ and } n \neq 0$$

which is true since for $m = 374 \in \mathbb{Z}$ and $n = 100 \in \mathbb{Z}$ we have $3.74 = \frac{374}{100}$. If we assign the value π to the variable x then we have the statement

$$\pi = \frac{m}{n} \text{ where } m, n \in \mathbb{Z}$$

which is false since there are no values $m, n \in \mathbb{Z}$ such that $\pi = \frac{m}{n}$.

Example 4.5

Examples of sets written in predicate form.

- $\{x \in \mathbb{N} \mid x \bmod 2 = 0\}$ = The set of positive even numbers.
- $\{x \in \mathbb{N} \mid x \bmod 2 = 1\}$ = The set of positive odd numbers.
- $\{x \in \mathbb{N} \mid 5 \leq x \leq 10\} = \{5, 6, 7, 8, 9, 10\}$
- $\mathbb{Q} = \left\{x \mid x = \frac{m}{n} \text{ for } m, n \in \mathbb{Z}, n \neq 0\right\}$ = The set of numbers $\frac{m}{n}$ where both m and n are elements of $\mathbb{Z}$ and $n \neq 0$
  = The set of all fractions.

Now consider the sets $\{1, 7, 3\}$ and $\{3, 7, 1\}$. These are both the same set so we will call them equal to each other. The order in which we write the elements does not matter at all. In fact, repeated elements do not matter either. Now consider the sets $\{1, 7, 3\}$ and $\{1, 3, 1, 7\}$. These are also the same set even though the element 1 was repeated twice in the second set.

Example 4.6

Examples of equal sets.

- $\{1, 7, 3\} = \{1, 3, 1, 7\}$

- $\{2, 4, 6, 8, 10\} = \{10, 8, 10, 6, 10, 4, 10, 4, 10, 2, 6, 8\}$

- $\{\frac{1}{3}, \frac{1}{2}, \frac{2}{3}\} = \{\frac{1}{3}, \frac{1}{3}, \frac{1}{3}, \frac{1}{2}, \frac{2}{3}, \frac{2}{3}, \frac{2}{3}\}$

- $\{\frac{1}{2}\} = \{\frac{1}{2}, \frac{2}{4}, \frac{3}{6}, \frac{4}{8}, \frac{5}{10}, \frac{6}{12}, \ldots\}$

- $\{\frac{1}{3}\} = \{\frac{1}{3}, \frac{2}{6}, \frac{3}{9}, \frac{4}{12}, \frac{5}{15}, \frac{6}{18}, \ldots\}$ There are many different ways to write any fraction.

There is one last, but very important, point that needs to be made. Sets can be elements of sets. This can cause a lot of confusion if you are not careful. Let us consider the following set,

$$S = \Big\{ a, b, c, \{b, c\}, \{a, b, c\}, d, \{a, d, h\}, h, i, \{h, i\} \Big\}.$$

Here we have put all the elements of the set in gray. These elements are separated by black commas. So we would say

$$a \in S, \quad b \in S, \quad c \in S, \quad d \in S, \quad h \in S, \quad \text{and} \quad i \in S.$$

But we would also say

$$\{b, c\} \in S, \quad \{a, b, c\} \in S, \quad \{a, d, h\} \in S, \quad \text{and} \quad \{h, i\} \in S.$$

These elements of S are actually sets themselves. This is allowable. Often in computer science there are situations where sets are considered elements of other sets. If you are given an example like this you must look very closely at the commas. We use commas to separate elements in a set. Here the commas that separate the elements of the set S are shown in black. But consider the element $\{b, c\}$ in S. It is a set that also uses commas to separate the elements b and c in it. In the set S this comma is in gray since it is part of the element $\{b, c\}$. Most of the time we will not write the elements of a set in gray so you must be careful and pay special attention to the commas.

4.2 SET OPERATIONS

Let A and B be sets. We say that B is a **subset** of A if every element of B is an element of A. If B is a subset of A we write $B \subseteq A$. If B contains an element that is not in A then B is not a subset of A and we write $B \nsubseteq A$. Notice that since set A contains every element of set A that A is a subset of A. That is, $A \subseteq A$. Clearly we have $A = A$ too. That is why $\subseteq$ has an extra line under the $\subset$, it shows that the sets could possibly be equal. However, if we have $B \subseteq A$ but $B \neq A$ then B is said to be a **proper subset** of A and we write $B \subset A$.

Example 4.7

Some examples of subsets.

- $\{2, 3, 5\} \subseteq \{1, 2, 3, 4, 5\}$.
  Furthermore, since $\{2, 3, 5\} \neq \{1, 2, 3, 4, 5\}$ then $\{2, 3, 5\}$ is a proper subset of $\{1, 2, 3, 4, 5\}$ and we can also write $\{2, 3, 5\} \subset \{1, 2, 3, 4, 5\}$.

- $\{2, 4, 6, 8\} \subseteq \{1, 2, 3, 4, 5, 6, 7, 8, 9, 10\}$.
  Again, since $\{2, 4, 6, 8\} \neq \{1, 2, 3, 4, 5, 6, 7, 8, 9, 10\}$ we have a proper subset and could also write $\{2, 4, 6, 8\} \subset \{1, 2, 3, 4, 5, 6, 7, 8, 9, 10\}$.

- $\{\,\} \subseteq \{2, 4, 6, 8, 10\}$.
  This is a very special subset. It is the empty subset. Clearly $\{\,\} \neq \{2, 4, 6, 8, 10\}$ so we could also write $\{\,\} \subset \{2, 4, 6, 8, 10\}$.

- $\{1, 2, 3, 4\} \subseteq \{1, 2, 3, 4\}$.
  Here, we have $\{1, 2, 3, 4\} = \{1, 2, 3, 4\}$ so we do not have a proper subset and we would write $\{1, 2, 3, 4\} \not\subset \{1, 2, 3, 4\}$.

- $\mathbb{N} \subseteq \mathbb{Z} \subseteq \mathbb{Q} \subseteq \mathbb{R}$.
  By looking at our definitions it is clear that we could also write $\mathbb{N} \subset \mathbb{Z} \subset \mathbb{Q} \subset \mathbb{R}$.

- If A is a set then $A \subseteq A$. Clearly we also have $A \not\subset A$ since $A = A$.

In the above example we saw the **empty set**, $\{\,\} \subseteq \{2, 4, 6, 8, 10\}$. The empty set is exactly that, a set that is empty. Notice we did not put any elements at all inside the curly-brackets. The curly-brackets are empty. Every set has the empty set as a subset. The empty set is so special it has its own symbol,

$$\varnothing = \{\,\}.$$

The empty set is also called the **null set**. Null is an old-fashioned word meaning nothing. It is interesting to notice that the empty set can be written in predicate form using a predicate that is always false,

$$\varnothing = \{x \mid x \neq x\}.$$

Clearly, there are no values of x that are not equal to themselves, thus this set has no values in it; it is empty.

Now let us return to the example we gave in the last section where some of the elements of the set S were also sets themselves,

$$S = \Big\{\, a\,,\ b\,,\ c\,,\ \{b, c\}\,,\ \{a, b, c\}\,,\ d\,,\ \{a, d, h\}\,,\ h\,,\ i\,,\ \{h, i\}\, \Big\}.$$

Again we have written the elements of S in gray. Pay close attention to the commas. Now we will write S with the elements b and c in gray, the element $\{\mathbf{b}, \mathbf{c}\}$ in gray boldface, and the rest of the elements in black,

$$S = \Big\{\, a\,,\ b\,,\ c\,,\ \{b, c\}\,,\ \{a, b, c\}\,,\ d\,,\ \{a, d, h\}\,,\ h\,,\ i\,,\ \{h, i\}\, \Big\}.$$

If we write $\{b, c\} \subset S$ then what we mean is

$$\{b, c\} \subset S.$$

In other words $\{b,c\}$ is the subset of S that contains the elements b and c written in gray. But if we write $\{\{b,c\}\} \subset S$ then what we mean is

$$\{\{\mathbf{b},\mathbf{c}\}\} \subset S.$$

In other words $\{\{b,c\}\}$ is the subset of S that contains the element $\{\mathbf{b},\mathbf{c}\}$ written in gray boldface. The subset $\{b,c,\{b,c\}\}$ would of course mean

$$\Big\{b,c,\{\mathbf{b},\mathbf{c}\}\Big\} \subset S$$

where b, c, and $\{b,c\}$ are all elements of S.

Example 4.8

Using the same set as before we give some examples of both elements and subsets to help you understand the difference;

$$S = \Big\{\, a\,,\, b\,,\, c\,,\, \{b,c\}\,,\, \{a,b,c\}\,,\, d\,,\, \{a,d,h\}\,,\, h\,,\, i\,,\, \{h,i\}\,\Big\}.$$

- The following are all elements in S;

$$a \in S, \quad b \in S, \quad c \in S, \quad d \in S, \quad h \in S, \quad i \in S.$$

- The following are all also elements in S;

$$\{b,c\} \in S, \quad \{a,b,c\} \in S, \quad \{a,d,h\} \in S, \quad \{h,i\} \in S.$$

- The following are all subsets of S that contain one element from S;

$$\{a\} \subset S, \quad \{b\} \subset S, \quad \{c\} \subset S, \quad \{d\} \subset S, \quad \{h\} \subset S, \quad \{i\} \subset S.$$

- The following are all also subsets of S that contain one element from S;

$$\{\{b,c\}\} \subset S, \quad \{\{a,b,c\}\} \subset S, \quad \{\{a,d,h\}\} \subset S, \quad \{\{h,i\}\} \subset S.$$

- The following are some examples of subsets of S that contain two elements;

$$\{a,b\} \subset S, \quad \{c,d\} \subset S, \quad \{a,\{b,c\}\} \subset S, \quad \{\{b,c\},\{a,d,h\}\} \subset S.$$

- The following are some examples of subsets of S that contain three elements;

$$\{c,d,i\} \subset S, \{h,i,\{h,i\}\} \subset S, \{d,\{b,c\},\ \{h,i\}\} \subset S, \{\{h,i\},\{a,d,h\},\{a,b,c\}\} \subset S.$$

There is another set that is very important in set theory called the **universal set**. The universal set is not the universe, but it is the "universe" of our problem. That means the universal set is the set that contains all the elements that arise in a particular problem.[2] So what we call the universal set for a problem depends on the problem. Each problem could have a different universal set. The universal set is usually represented by $\mathcal{U}$. We will use the universal set in the next section.

---

[2]Much later in your computer science major you may see problems where there is no well-defined universal set.

The **intersection** of two sets A and B is given by

$$A \cap B = \{x \mid x \in A \text{ and } x \in B\}.$$

That is, the intersection of two sets A and B consists of the set of elements that are in set A and also in set B. Another way of saying this is that the intersection of two sets consists of the elements that are common to both sets.

Example 4.9

Intersections of sets.

- If $A = \{1, 2, 3, 4, 5\}$ and $B = \{4, 5, 6, 7\}$ then $A \cap B = \{4, 5\}$.
- If $A = \{1, 2, 3, 4, 5, 6, 7, 8\}$ and $B = \{2, 4, 6, 8, 10, 12\}$ then $A \cap B = \{2, 4, 6, 8\}$.
- If $A = \{2, 4, 6, 8, 10, 12, \ldots\}$ and $B = \{1, 3, 5, 7, 9, 11, \ldots\}$ then $A \cap B = \{\,\} = \varnothing$.

In the above example when A was the set of even numbers and B was the set of odd numbers we saw that $A \cap B = \varnothing$. In other words, there are no elements common to the set of even numbers and the set of odd numbers. A number simply cannot be both even and odd at the same time. These sets are called disjoint. If A and B are sets and $A \cap B = \varnothing$ then A and B are said to be **disjoint**.

The **union** of two sets A and B is given by

$$A \cup B = \{x \mid x \in A \text{ or } x \in B\}.$$

That is, the union of two sets A and B consists of all the elements in A together with all the elements in B.

Example 4.10

Unions of sets.

- If $A = \{1, 2, 3, 4, 5\}$ and $B = \{4, 5, 6, 7\}$ then $A \cup B = \{1, 2, 3, 4, 5, 6, 7\}$.
- If $A = \{1, 2, 3, 4, 5, 6, 7, 8\}$ and $B = \{2, 4, 6, 8, 10, 12\}$ then $A \cup B = \{1, 2, 3, 4, 5, 6, 7, 8, 10, 12\}$.
- If $A = \{2, 4, 6, 8, 10, 12, \ldots\}$ and $B = \{1, 3, 5, 7, 9, 11, \ldots\}$ then $A \cup B = \mathbb{N}$.

The **complement** of a set A is the set of all elements in the universal set but not in A. The complement of A is written as A^c or as $\bar{A}$ or as A'. In order to find the complement of a set you need to know what the universal set $\mathcal{U}$ is.

Example 4.11

Complements of sets.

- If $\mathcal{U} = \{1, 2, 3, 4, 5, 6\}$ and $A = \{2, 4, 6\}$ then $A^c = \{1, 3, 5\}$.
- If $\mathcal{U} = \{1, 2, 3, 4, 5, 6, 7, 8, 9, 10\}$ and $A = \{1, 2, 3, 9, 10\}$ then $A^c = \{4, 5, 6, 7, 8\}$.
- If $\mathcal{U} = \mathbb{N}$ and $A = \{2, 4, 6, 8, 10, 12, \ldots\}$ then $A^c = \{1, 3, 5, 7, 9, 11, \ldots\}$.

The **difference** between sets A and B is given by

$$A - B = \{x \mid x \in A \text{ and } x \notin B\}.$$

That is, $A - B$ is the set of elements in A that are not also in B.

Example 4.12

Differences of sets.

- If $A = \{1, 2, 3, 4, 5, 8, 10\}$ and $B = \{1, 5, 6, 7, 10\}$ then $A - B = \{2, 3, 4, 8\}$.
- If $A = \{4, 5, 6, 7, 8, 9\}$ and $B = \{1, 3, 5, 7, 9, 11\}$ then $A - B = \{4, 6, 8\}$.
- If $A = \{2, 4, 6, 8, 10, 12, \ldots\}$ and $B = \{1, 3, 5, 7, 9, 11, \ldots\}$ then $A - B = A$.

The **cardinality** of a finite set is the number of elements in the set. If A is a set the cardinality of A is written as $|A|$ or as $n(A)$. If the set has an infinite number of elements then we say the cardinality of the set is infinite. Since the empty set has no elements then the cardinality of the empty set is zero.

Example 4.13

Cardinality of sets.

- If $A = \{1, 2, 3, 4, 5, 6, 7, 8, 9, 10\}$ then $|A| = 10$.
- If $A = \{2, 4, 6, 8, 10, 12, \ldots\}$ then $|A| = \infty$, that is, $|A|$ is infinite.
- If $A = \{\ \} = \varnothing$ then $|A| = |\varnothing| = 0$.

The **power set** of a set A is the set that contains all the subsets of set A. The power set of A is written as $\mathcal{P}(A)$. Notice, here we are talking about a set that actually has sets as its elements. That is entirely possible.

Example 4.14

We will find the power set of the set $A = \{a, b, c\}$. Notice that A has three elements, the letters a, b, and c. We list the subsets of A:

$$\varnothing, \{a\}, \{b\}, \{c\}, \{a, b\}, \{a, c\}, \{b, c\}, \{a, b, c\}.$$

Notice that the empty set is a subset of A and the set A is a subsets of itself. These eight sets are the elements of the power set of A. The power set of A is given by

$$\mathcal{P}(A) = \Big\{\varnothing, \{a\}, \{b\}, \{c\}, \{a, b\}, \{a, c\}, \{b, c\}, \{a, b, c\}\Big\}.$$

Given a subset of A, for each element in A there are two possibilities, either that the element is in the subset or it is not in the subset. Let us look at a few more examples to see how this works.

Example 4.15

Cardinality of power sets.

- Let $A = \{a\}$. The set A only has one element. There are two possible subsets, one that has the element a in it, $\{a\}$, and one that does not have the element a in it, $\{\ \}$. So A has $2^1 = 2$ subsets and $\mathcal{P}(A) = \{\varnothing, \{a\}\}$.
- $A = \{a, b\}$. The set A has two elements in it. For each element there are two possibilities, either it is in the set or it is not in the set. Thus we have

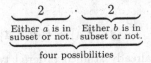

which give a total of $2 \cdot 2 = 4$ possibilities. That means there are $2^2 = 4$ subsets, $\mathcal{P}(A) = \{\{\ \}, \{a\}, \{b\}, \{a, b\}\}$.
- $A = \{a, b, c\}$. The set A has three elements in it. For each element there are two possibilities, either it is in the set or it is not in the set. Thus we have

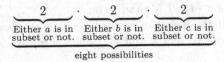

which give a total of $2 \cdot 2 \cdot 2 = 8$ possibilities. This means A has $2^3 = 8$ subsets. The power set is given in the last example.

By now the pattern should be clear. This leads to the following theorem.

Theorem 4.1 *If* $|A| = n$ *then* $|\mathcal{P}(A)| = 2^n$.

Example 4.16

Supposes $A = \{a, b, c, d\}$. How many subsets does A have? How many proper subsets does A have?

The power set is the set that contains all the subsets of A, so the cardinality of the power set is equal to the number of subsets that A has. Since $|A| = 4$ then $|\mathcal{P}(A)| = 2^4 = 16$. Thus A has 16 subsets. However, one of these subsets is A itself, so there are only 15 proper subsets of A.

Example 4.17

Suppose that A has 127 proper subsets. How many elements does A have?

Since A has 127 proper subsets it must have 128 subsets in all. This is because there is one subset that is not a proper subset, namely A itself, which must be added to 127. This means that $|\mathcal{P}(A)| = 128$. We now need to find the power of two that is equal to 128. Since $2^7 = 128$ we have $|A| = 7$. In other words, A has seven elements.

Recall, that in sets the order we write the elements does not matter. Thus $\{a, b, c\}$ is exactly the same as $\{b, c, a\}$ or $\{a, c, b\}$ or $\{b, a, c\}$ and so on. But there are many cases in computer science where order is very important. We want a way to think about things that have an order. An **ordered n-tuple** is a list of n elements in a particular order. Ordered n-tuples are written in parenthesis instead of curly-brackets to distinguish them from sets. So $(2, 3)$ is an ordered two-tuple while $\{2, 3\}$ is a set and $(3, -2, 4)$ is an ordered three-tuple while $\{3, -2, 4\}$ is a set. But while the order of elements in a set does not matter, the order of the terms in an ordered n-tuple does matter. Thus,

$$(3, -2, 4) \neq (-2, 3, 4)$$

but

$$\{3, -2, 4\} = \{-2, 3, 4\}.$$

In fact, you have doubtless already studied ordered two-tuples and ordered three-tuples when you studied points in two or three dimensions. Points in n dimensions are nothing more than ordered n-tuples of numbers. See Figs. 4.1 and 4.2 that show how points are n-tuples. But of course the order we use when writing down a point is very important. In the plane $\mathbb{R}^2$ the point $(2, -1)$ is very different from the point $(-1, 2)$. Similarly, in three-dimensional space $\mathbb{R}^3$ the point $(-2, 5, 4)$ is very different from the point $(5, -2, 4)$, which is different from the point $(4, -2, 5)$, and so on.

Can we have a set that consists of ordered n-tuples as elements? Of course we can. The **Cartesian product**[3] of two sets A and B is defined to be

$$A \times B = \big\{ (x, y) \mid x \in A \text{ and } y \in B \big\}.$$

That is, $A \times B$ is the set of all ordered pairs where the first term in the ordered pair comes from set A and the second term in the ordered pair comes from B. Similarly, the Cartesian product of three sets A, B, and C is defined to be

$$A \times B \times C = \big\{ (x, y, z) \mid x \in A \text{ and } y \in B \text{ and } z \in C \big\}.$$

That is, $A \times B \times C$ is the set of all ordered three-tuples where the first term in the ordered three-tuple comes from set A, the second term in the ordered three-tuple comes from B, and the third term in the ordered three-tuple comes from C. In general we have

$$S_1 \times S_2 \times \cdots \times S_n = \big\{ (x_1, x_2, \ldots, x_n) \mid x_1 \in S_1, x_2 \in S_2, \cdots, x_n \in S_n \big\}.$$

Example 4.18

Cartesian products of sets.

- Let $A = \{a, b, c\}$ and let $B = \{2, 4\}$. Then
$$A \times B = \big\{ (a, 2), (b, 2), (c, 2), (a, 4), (b, 4), (c, 4) \big\}.$$

- The set of points in the plane are given by
$$\mathbb{R} \times \mathbb{R} = \big\{ (x, y) \mid x \in \mathbb{R} \text{ and } y \in \mathbb{R} \big\}.$$

In fact, this explains why the plane is often written as $\mathbb{R}^2$, since using the rule of exponents we can write $\mathbb{R}^2 = \mathbb{R} \times \mathbb{R}$. See Fig. 4.1.

---

[3]The Cartesian product is named after Cartesius, the Latinized version of the last name of the French philosopher René Descartes. Since it is named after a person often it is capitalized, but sometimes it is not. Both conventions are considered grammatically correct.

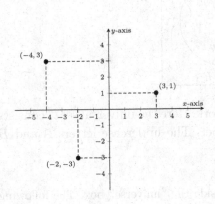

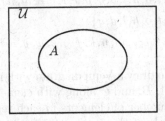

Figure 4.1 Points in the two-dimensional plane $\mathbb{R}^2$ are ordered two-tuples.

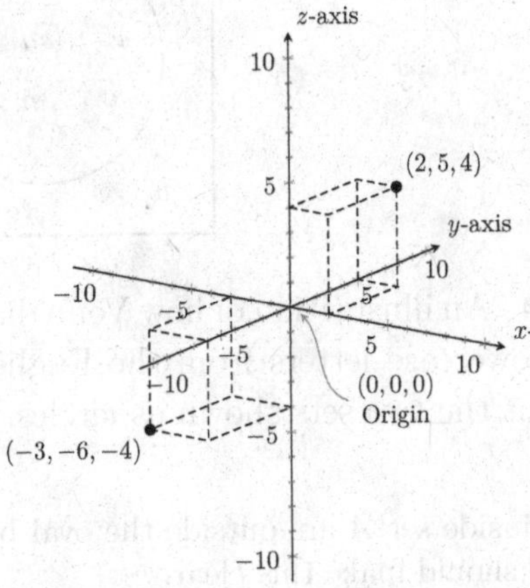

Figure 4.2 Points in three-dimensional space $\mathbb{R}^3$ are ordered three-tuples.

- The set of points in three-dimensional space are given by

$$\mathbb{R} \times \mathbb{R} \times \mathbb{R} = \big\{ (x, y, z) \mid x \in \mathbb{R}, y \in \mathbb{R} \text{ and } z \in \mathbb{R} \big\}.$$

In fact, this explains why three-dimensional space is often written as $\mathbb{R}^3$, since using the rule of exponents we can write $\mathbb{R}^3 = \mathbb{R} \times \mathbb{R} \times \mathbb{R}$. See Fig. 4.2.

- Let $L = \{a, b, c, d, \ldots, x, y, z\}$. The set of all strings of four letters is given by $L \times L \times L \times L$.

4.3 VENN DIAGRAMS

Venn diagrams are very common in set theory. They are a very powerful tool to help you visualize what is happening in set theory. But be careful! Venn diagrams are only a picture, or cartoon, that helps us understand what is going on. They are not always accurate. In fact, at times they can be misleading. However, they are very helpful to understand what set operations do. We will use Venn diagrams to show pictures of some of the set operations from the last section.

Figure 4.3 The universal set $\mathcal{U}$ and the set A.

Look at Fig. 4.3. Venn diagrams always include the universal set which is usually shown as a box that surrounds everything else and is often labeled $\mathcal{U}$. However, very often for the universal set one does not bother to write the label $\mathcal{U}$ down. The set A is shown as the oval. We imagine the elements of set A as being inside the oval labeled A. The elements that

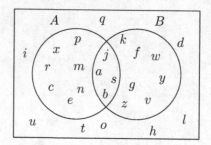

Figure 4.4 An illustration of how Venn diagrams work. Here the universal set $\mathcal{U}$ consists of the lowercase letters from the English alphabet. The uppercase letters A and B represent the two sets shown as circles.

are not inside set A are outside the oval but still inside the "universe" box. The following example should make this clear.

Example 4.19

In Fig. 4.4 we have the universal set $\mathcal{U}$ which consists of all the lowercase letters from the English alphabet. There are also two sets in $\mathcal{U}$, set A and set B. The letters in set A are shown inside the circle labeled A and the elements in set B are shown inside circle B. Find sets A and B, and then find $A \cap B$, $A \cup B$, A^c, B^c, $(A \cup B)^c$, and $(A \cap B)^c$.

We find sets A and B by listing all the elements that are in the circle labeled A and in the circle labeled B.

$$A = \{p, x, r, c, e, m, n, j, a, s, b\},$$
$$B = \{j, a, s, b, k, f, g, z, w, y, v\}.$$

We find the sets $A \cap B$, $A \cup B$, A^c, B^c, $(A \cup B)^c$, and $(A \cap B)^c$ by listing all the elements in the appropriate region.

$$A \cap B = \{j, a, s, b\},$$
$$A \cup B = \{p, x, r, c, e, m, n, j, a, s, b, k, f, g, z, w, y, v\},$$
$$A^c = \{i, u, t, o, h, l, d, q, k, f, g, z, w, y, v\},$$
$$B^c = \{i, u, t, o, h, l, d, q, p, x, r, c, e, m, n\},$$
$$(A \cup B)^c = \{i, u, t, o, h, l, d, q\},$$
$$(A \cap B)^c = \{p, x, r, c, e, m, n, k, f, g, z, w, y, v, i, u, t, o, h, l, d, q\}.$$

It is also of course possible to draw a Venn diagram with more than two sets. Fig. 4.5 is a Venn diagram with three sets A, B, and C along with elements shown. A lot of times Venn diagrams are drawn with the number of elements in each region instead of with the actual elements themselves. An example of this is given in Fig. 4.6. When we do this we do not know what the elements are and so we cannot write out the sets A, B, or C in enumerated form. However, often we are more interested in simply knowing how many elements are in each set. In other words, we only care about the cardinalities of the set. We can still use Venn diagrams for this.

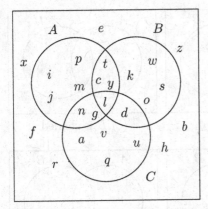

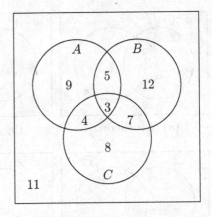

Figure 4.5 A Venn diagram with three sets, A, B, and C, and with elements shown.

Figure 4.6 In this Venn diagram the number of elements in each region is indicted, not the actual elements themselves.

Example 4.20

Using the Venn diagram in Fig. 4.6 find $|A|$, $|B|$, $|C|$, $|A\cap B|$, $|A\cap C|$, $|B\cap C|$, $|A\cap B\cap C|$, $|A\cup B|$, $|A\cup C|$, $|B\cup C|$, $|A\cup B\cup C|$, $|A^c|$, and $|(A\cup B\cup C)^c|$.

$$|A| = 9 + 5 + 3 + 4 = 21$$
$$|B| = 12 + 5 + 3 + 7 = 27,$$
$$|C| = 8 + 7 + 3 + 4 = 22,$$
$$|A\cap B| = 5 + 3 = 8,$$
$$|A\cap C| = 4 + 3 = 7,$$
$$|B\cap C| = 7 + 3 = 10,$$
$$|A\cap B\cap C| = 3,$$
$$|A\cup B| = 9 + 4 + 5 + 3 + 12 + 7 = 40,$$
$$|A\cup C| = 9 + 5 + 4 + 3 + 8 + 7 = 36,$$
$$|B\cup C| = 12 + 5 + 7 + 3 + 8 + 4 = 39,$$
$$|A\cup B\cup C| = 9 + 5 + 12 + 4 + 3 + 7 + 8 = 48,$$
$$|A^c| = 12 + 7 + 8 + 11 = 38,$$
$$|(A\cup B\cup C)^c| = 11.$$

In a general situation we often use shading to indicate the elements in a set. In Fig. 4.7(a) the set $\mathcal{U}$ is shaded which indicates we are considering all the elements in the "universe." In Fig. 4.7(b) the set A is shaded which indicates we are considering all the elements in the set A. In Fig 4.7(c) the set B is shaded indicating we are considering all the elements in set B. Figs. 4.7(d) through 4.7(k) show all the other possibilities as well. These pictures are a very nice way to visualize what union and intersection and complement all mean. Be careful though, with more then three sets Venn diagrams can start to be misleading and confusing.

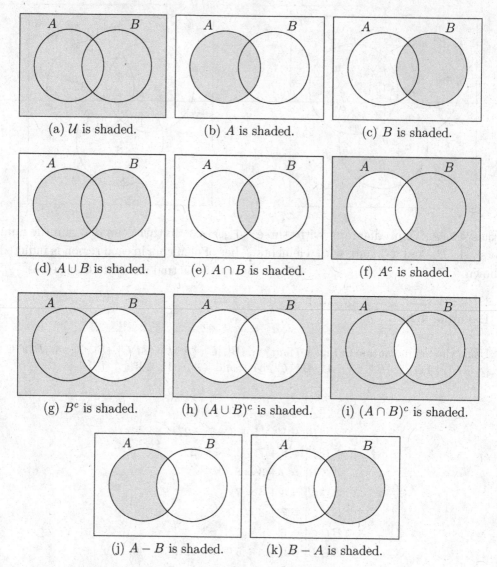

Figure 4.7 Using shading to indicate various regions in Venn diagrams.

4.4 THE LAWS OF SET THEORY

Just like there are laws of logic, there are also **laws of set theory**. If you compare the laws of logic with the laws of set theory you will see that they are basically the same. That is because the set operations of intersection, union, and complement correspond in a natural way to the logical connectives **and, or**, and **not**. The laws of logic could be used to simplify complicated logical expressions. Similarly, the laws of set theory can be used to simplify complicated set theory expressions. Given a universal set $\mathcal{U}$, the set operations $\cap$, $\cup$, and ^{c} operate on subsets of the universal set. Thus we say that $\cap$, $\cup$, and ^{c} operate on elements of $\mathcal{P}(\mathcal{U})$, the powerset of $\mathcal{U}$.

Venn diagrams can be very useful in helping us understand the laws of set theory. But be careful, Venn diagrams can be used to understand the laws of set theory, but they cannot be used to actually prove the laws of set theory. To prove the laws of set theory you need to use the actual definitions of the operations.

Name	Laws of Set Theory	
Commutative laws	(a) $A \cap B = B \cap A$	(b) $A \cup B = B \cup A$
Associative laws	(a) $(A \cap B) \cap C = A \cap (B \cap C)$	(b) $(A \cup B) \cup C = A \cup (B \cup C)$
Distributive laws	(a) $A \cap (B \cup C) = (A \cap B) \cup (A \cap C)$	(b) $A \cup (B \cap C) = (A \cup B) \cap (A \cup C)$
Identity laws	(a) $A \cap \mathcal{U} = A$	(b) $A \cup \emptyset = A$
Inverse laws	(a) $A \cap A^c = \emptyset$	(b) $A \cup A^c = \mathcal{U}$
Double complement law	$(A^c)^c = A$	
Idempotent laws	(a) $A \cap A = A$	(b) $A \cup A = A$
DeMorgan's laws	(a) $(A \cap B)^c = A^c \cup B^c$	(b) $(A \cup B)^c = A^c \cap B^c$
Annihilation laws	(a) $A \cap \emptyset = \emptyset$	(b) $A \cup \mathcal{U} = \mathcal{U}$
Absorption laws	(a) $A \cap (A \cup B) = A$	(b) $A \cup (A \cap B) = A$
Complement laws	(a) $\mathcal{U}^c = \emptyset$	(b) $\emptyset^c = \mathcal{U}$

Example 4.21

Use Venn diagrams to argue that the double complement law is true.

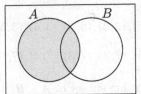 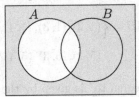

On the left the set A is shown. On the right we see A^c. But just by looking at these pictures it is obvious that $(A^c)^c$ must be the same as A.

Example 4.22

Use Venn diagrams to argue the first of DeMorgan's laws, $(A \cap B)^c = A^c \cup B^c$

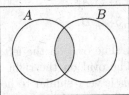 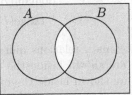

On the left we see set $A \cap B$ and on the right we see $(A \cap B)^c$.

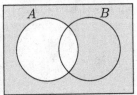 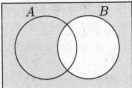 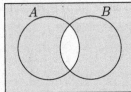

On the left we show A^c. In the middle we show B^c. On the right we see the union of these two sets, $A^c \cup B^c$.

4.5   BINARY RELATIONS ON SETS

In computer science relations on sets are important when working with databases. We use the language of set theory when discussing relations. Suppose A and B are sets. A **binary relation** from set A to set B is a subset of the cartesian product $A \times B$. Usually this subset is called R. The word binary is used to indicate that there are two sets involved, sets A and B. If we were to let sets $A = B$ then we would say that a binary relation on set A is a subset of $A \times A$. Very often a binary relation is simply called a **relation**. We use some of the same terms from function theory. The domain and range of a relation R are given by

$$\text{domain}(R) = \{x \mid (x, y) \in R\},$$
$$\text{range}(R) = \{y \mid (x, y) \in R\}.$$

If $(x, y) \in R$ then we often write xRy. This means that x is related to y.

Example 4.23

Let $A = \{1, 3, 5, 7\}$ and $B = \{2, 3, 4, 5\}$. First we find the Cartesian product of sets A and B,

$$A \times B = \Big\{(1,2), (1,3), (1,4), (1,5), (3,2), (3,3), (3,4), (3,5),$$
$$(5,2), (5,3), (5,4), (5,5), (7,2), (7,3), (7,4), (7,5)\Big\}.$$

The relation "is equal to" is given by the following subset of $A \times B$,

$$R = \{(3,3), (5,5)\} \subset A \times B.$$

Since $(3,3) \in R$ then we would write $3R3$ and would read it out loud as "3-are-3." Since the relation is "is equal to" then $3R3$ could also be read out loud as "3 is equal to 3." Also, since $(5,5) \in R$ then we would write that as $5R5$ and read it as "5-are-5" or as "5 is equal to 5." We also have

$$\text{domain}(R) = \{3, 5\} \subset A,$$
$$\text{range}(R) = \{3, 5\} \subset B.$$

We can also draw binary relations quite easily. The oval on the left represents set A and its elements are shown inside the oval. The oval on the right represents set B and its elements are shown inside the oval. This is very similar to how we draw Venn diagrams. The relation is represented as arrows between the elements of set A and set B. For example, since $(3,3) \in R \subset A \times B$ then there is an arrow from $3 \in A$ to $3 \in B$. And since $(5,5) \in R \subset A \times B$ then there is another arrow from $5 \in A$ to $5 \in B$.

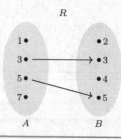

Example 4.24

Let $A = \{1,3,5,7\}$ and $B = \{2,3,4,5\}$. The Cartesian product $A \times B$ was found in the last example. The relation "is less than" is given by the following subset of $A \times B$,

$$R = \{(1,2),(1,3),(1,4),(1,5),(3,4),(3,5)\} \subset A \times B.$$

Since $(1,2) \in R$ then we can write $1R2$ which can be read "1-are-2" or "1 is less than 2." Similarly, $(1,3) \in R$ can be written as $1R3$ or read as "1-are-3" or as "1 is less than 3," and so on. We also have

$$\text{domain}(R) = \{1,3\} \subset A,$$
$$\text{range}(R) = \{2,3,4,5\} \subset B.$$

This relation is also shown in the following picture. Since $(1,2),(1,3),(1,4),(1,5) \in R \subset A \times B$ then we have arrows going from $1 \in A$ to $2,3,4,5 \in B$. Since $(3,4),(3,5) \in R \subset A \times B$ then we have arrows going from $3 \in A$ to $4,5 \in B$.

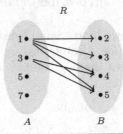

Example 4.25

Let $A = \{1,3,5,7\}$ and $B = \{2,3,4,5\}$. The relation "is greater than" is given by the following subset of the Cartesian product $A \times B$,

$$R = \{(3,2),(5,2),(5,3),(5,4),(7,2),(7,3),(7,4),(7,5)\} \subset A \times B.$$

Since $(3,2) \in R$ then we can write $3R2$ which can be read "3-are-2" or "3 is greater than 2." Similarly, $(5,2) \in R$ can be written as $5R2$ and read as "5-are-2" or as "5 is greater than 2," and so on. We also have

$$\text{domain}(R) = \{3,5,7\} \subset A,$$
$$\text{range}(R) = \{2,3,4,5\} \subset B.$$

This relation is also shown in the following picture,

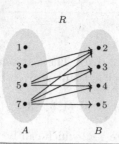

R

A   *B*

When working with databases we may have relations that are not related to mathematics. Here are some simple examples of relations that one could encounter when working with databases.

Example 4.26

Suppose $A = \{Sam, Bob, Tim\}$ and $B = \{Fido, Hunter, Rex\}$. The relation OwnsDog is given by the following subset of the Cartesian product $A \times B$,

$$\text{OwnsDog} = \Big\{(Sam, Fido), (Bob, Rex), (Tim, Hunter)\Big\} \subset A \times B.$$

Since $(Sam, Fido) \in$ OwnsDog we have "Sam OwnsDog Fido" which clearly means that Sam owns the dog named Fido. "Bob OwnsDog Rex" means that Bob owns the dog Rex. And similarly Tim owns the dog Hunter. The relation can be pictured as below.

OwnsDog

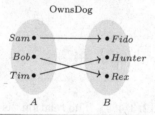

A   *B*

Example 4.27

Suppose $A = \{John, Sally, Anne, Ted, Bill, Ed, Jane\}$. We have a relation called ChildOf given by the following subset of the Cartesian product $A \times A$,

$$\text{ChildOf} = \Big\{(John, Sally), (Ted, Bill), (John, Ted),$$

$$(Bill, Ed), (Bill, Anne), (Ted, Jane)\Big\} \subset A \times A.$$

Since we have $(John, Sally) \in$ ChildOf we have "John ChildOf Sally" which is easy to understand means that John is a child of Sally. "Ted ChildOf Bill" of course means Ted is a child of Bill, and so on. Can you draw a family tree based on this relation?

When R is a binary relation on a finite set A, that is, when $R \subset A \times A$, then there is another common way to visualize the relation R as a directed graph. The set A becomes the vertex set of the directed graph and the elements $(x, y) \in R$ are used to make the edge set. The element (x, y) indicates a directed edge from vertex x to vertex y. We will discuss directed graphs more in chapter 9, but for now we just want to point out the connection between set theory and graph theory.

Example 4.28

Suppose $A = \{1, 2, 3\}$ and we have the relation

$$R = \big\{(1,1), (1,2), (3,2)\big\} \subset A \times A.$$

The relation R can be depicted as a directed graph by

We interpret $(1,1) \in R$ as meaning one is sent to one, we interpret $(1,2) \in R$ as meaning one is sent to two, and we interpret $(3,2) \in R$ as meaning three is sent to two.

Example 4.29

Functions can be defined in terms of a binary relation. We will consider this in depth in chapter 6, but for now we just want to point out the connection. Consider the function f shown below.

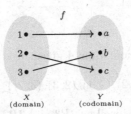

The domain is given by $X = \{1, 2, 3\}$ and the codomain is given by $Y = \{a, b, c\}$. The Cartesian product of X and Y is given by

$$X \times Y = \big\{(x, y) \mid x \in X, y \in Y\big\}$$
$$= \big\{(1, a), (2, a), (3, a), (1, b), (2, b), (3, b), (1, c), (2, c), (3, c)\big\}.$$

The function f is a subset of $X \times Y$ given by

$$f = \big\{(1, a), (2, c), (3, b)\big\} \subset X \times Y.$$

Functions are a special kind of binary relation that satisfy two extra conditions:

1. If $x \in X$ then there exists a $y \in Y$ such that $(x, y) \in f$,
2. and this $y \in Y$ is unique. (That means if (x, y) and (x, z) are in f then $y = z$.)

There are certain types of relations that are important in many applications. Suppose that A is a set and R is a relation on $A \times A$. If $(x, y) \in R$ then we write xRy.

1. R is called **reflexive** if xRx for every $x \in A$.
2. R is called **irreflexive** if there are no elements $x \in A$ such that xRx.

3. R is called **symmetric** if xRy implies yRx for all $x, y \in A$.
4. R is called **antisymmetric** if xRy and yRx implies $x = y$ for all $x, y \in A$.
5. R is called **transitive** if xRy and yRz implies xRz for all $x, y, z \in A$.

Example 4.30

Let $A = \{a, b, c\}$ with relation R on A given by

$$R = \big\{(a, a), (b, b), (c, c)\big\} \subset A \times A.$$

This means we have aRa, bRb, and cRc. In other words, we have xRx for every $x \in A$, which means that R is a reflexive relation.

Example 4.31

Let $A = \{a, b, c\}$ with relation R on A given by

$$R = \big\{(a, b), (b, c), (c, a)\big\} \subset A \times A.$$

This means we have aRb, bRc, and cRa. It is clear by looking at these that there is no element $x \in A$ such that xRx, which means that R is an irreflexive relation.

Example 4.32

Let $A = \{1, 2, 3\}$ with the relation R on A given by

$$R = \big\{(1, 2), (2, 1), (2, 3), (3, 2)\big\} \subset A \times A.$$

Clearly we have $1R2$ and $2R1$ as well as $2R3$ and $3R2$. We can see that if xRy is in R then yRx is also in R. Thus R is a symmetric relation.

Example 4.33

Let us consider the set $\mathbb{N}$ of natural numbers with the relation R given by

$$R = \big\{(x, y)\,|\,x \leq y\big\} \subset \mathbb{N} \times \mathbb{N}.$$

The relation R is also known as "is less than or equal to." For example, $(1, 2) \in R$ since $1 \leq 2$. But also $(1, 1) \in R$ since $1 \leq 1$. Suppose that for $x, y \in \mathbb{N}$ we have both xRy and yRx. In other words, we have that $x \leq y$ and $y \leq x$. The only way this can happen is if $x = y$. This means that R is an antisymmetric relation.

Example 4.34

Let $A = \{1, 2, 3\}$ with the relation R given by

$$R = \{(1,2), (2,3), (1,3)\} \subset A \times A.$$

We have $1R2$ and $2R3$ and also $1R3$. Thus this is a transitive relation.

There are two more relations that are extremely important in computer science and mathematics. These relations are defined using the relations listed above.

1. A relation is called an **equivalence relation** if it is reflexive, symmetric, and transitive.
2. A relation is called a **partial ordering** if it is reflexive, antisymmetric, and transitive.

Example 4.35

Let $A = \{1, 2, 3, 4, 5, 6\}$ with the relation given by

$$R = \{(1,1), (1,3), (1,5), (3,1), (3,3), (3,5), (5,1),$$
$$(5,3), (5,5), (2,2), (2,6), (6,2), (6,6), (4,4)\} \subset A \times A.$$

To check that R is reflexive we need to check that xRx for every $x \in A$. This is exactly

$$(1,1), (2,2), (3,3), (4,4), (5,5), (6,6) \in R.$$

Thus R is reflexive. To check that R is symmetric we must check that if xRy then yRx. This is exactly

$$(1,3), (3,1) \in R, \qquad\qquad (1,5), (5,1) \in R,$$
$$(3,5), (5,3) \in R, \qquad\qquad (2,6), (6,2) \in R.$$

Thus R is symmetric. To check that R is transitive we must check that if xRy and yRz then xRz. This is exactly

$$(1,3), (3,5), (1,5) \in R.$$

Thus R is transitive. Therefore R is an equivalence relation on A.

Example 4.36

The relation $\leq$ defined on $\mathbb{N}$ is a partial ordering on $\mathbb{N}$. We have $x \leq x$ for any $x \in \mathbb{N}$ so $\leq$ is reflexive. For every $x, y \in \mathbb{N}$ we have that if $x \leq y$ and $y \leq x$ then $x = y$. This makes $\leq$ antisymmetric. And for any $x, y, z \in \mathbb{N}$ we have that if $x \leq y$ and $y \leq z$ then $x \leq z$. This makes $\leq$ transitive. Since $\leq$ is reflexive, antisymmetric, and transitive then it is a partial ordering.

If A is a set then a **partition** of A is a set of subsets of A such that every element of A is in exactly one of the subsets. If R is an equivalence relation on A then for each $x \in A$ we can define a subset

$$E(x) = \{y \in A \,|\, yRx\}.$$

The sets $E(x)$ are called the **equivalence classes of** R and the set $E(x)$ is called the **equivalence class of** x. The set of equivalence classes of R is a partition of the set A. Thus equivalence relations can be used to partition a set into subsets called equivalence classes.

Example 4.37

Let $A = \{1, 2, 3, 4, 5, 6\}$ with the relation given by

$$R = \{(1,1), (1,3), (1,5), (3,1), (3,3), (3,5), (5,1),$$
$$(5,3), (5,5), (2,2), (2,6), (6,2), (6,6), (4,4)\} \subset A \times A.$$

Using the definition of equivalence class we find that

$$E(1) = \{1, 3, 5\},$$
$$E(2) = \{2, 6\},$$
$$E(3) = \{1, 3, 5\},$$
$$E(4) = \{4\},$$
$$E(5) = \{1, 3, 5\},$$
$$E(6) = \{2, 6\}.$$

Clearly $E(1) = E(3) = E(5)$ and $E(2) = E(6)$. Thus we have partitioned set A into three equivalence classes,

$$A = \{1, 3, 5\} \cup \{2, 6\} \cup \{4\}.$$

4.6 PROBLEMS

Question 4.1 *Let $A = \{2, 4, 8, 6, a, e, g, h, b, e\}$ and $B = \{6, 9, 2, 8, r, v, x, z, t, s\}$. Determine if the following statements are True or False.*

(a) $4 \in A$	(g) $7 \in B$	(m) $b \notin A$	(s) $5 \notin B$
(b) $6 \in A$	(h) $9 \in B$	(n) $8 \notin A$	(t) $s \notin B$
(c) $r \in A$	(i) $z \in B$	(o) $g \notin A$	(u) $u \notin B$
(d) $b \in A$	(j) $q \in B$	(p) $c \notin A$	(v) $8 \notin B$
(e) $f \in A$	(k) $8 \in B$	(q) $5 \notin A$	(w) $5 \notin B$
(f) $1 \in A$	(l) $y \in B$	(r) $3 \notin A$	(x) $v \notin B$

Question 4.2 *Let $A = \{2, 4, 8, 6, a, e, g, h, b\}$ and $B = \{6, 9, 2, 8, r, v, x, z, t, s\}$. Determine if the following statements are True or False.*

(a) $\{6, a, e\} \subset A$	(g) $\{9, 8, 7\} \subset B$	(m) $\{2, 4, 6, 8\} \not\subset A$	(s) $\{6, 8, t\} \not\subset B$
(b) $\{2, 4, 6, 8\} \subset A$	(h) $\{v, x, 8, 9\} \subset B$	(n) $\{a, e, i\} \not\subset A$	(t) $\{x, y, z\} \not\subset B$
(c) $\{g, 5, 6, j\} \subset A$	(i) $\{v\} \subset B$	(o) $\{a, b\} \not\subset A$	(u) $\{8, 6\} \not\subset B$
(d) $\{b, a, c\} \subset A$	(j) $\{\,\} \subset B$	(p) $\{4, h, 2\} \not\subset A$	(v) $\{\,\} \not\subset B$
(e) $\{a, 8\} \subset A$	(k) $\{3, 6, 9\} \subset B$	(q) $\{\,\} \not\subset A$	(w) $\{v, r, 8, 2\} \not\subset B$
(f) $\{\,\} \subset A$	(l) $\{z, 2\} \subset B$	(r) $\{9\} \not\subset A$	(x) $\{2, y, 3\} \not\subset B$

Question 4.3 Let $\mathcal{U} = \left\{1, 2, 3, a, b, c, \{a, b\}, \{1, 3\}, \{1, a, 3, c\}\right\}$. Determine if the following statements are True or False.

(a) $2 \in \mathcal{U}$

(b) $c \in \mathcal{U}$

(c) $\{1, 2\} \in \mathcal{U}$

(d) $\{1, 3\} \in \mathcal{U}$

(e) $d \in \mathcal{U}$

(f) $\{b, a\} \in \mathcal{U}$

(g) $\{b, c\} \in \mathcal{U}$

(h) $\{3, c, 1, a\} \in \mathcal{U}$

(i) $b \in \mathcal{U}$

(j) $a \in \mathcal{U}$

(k) $\{1, 2\} \subset \mathcal{U}$

(l) $\{1, 3\} \subset \mathcal{U}$

(m) $\{1, 2, 3\} \subset \mathcal{U}$

(n) $\{2, \{3, 1\}\} \subset \mathcal{U}$

(o) $\{\{a, b, c\}, a, b\} \subset \mathcal{U}$

(p) $\{1, a, 3, c\} \subset \mathcal{U}$

(q) $\{\{1, a, 3, c\}\} \subset \mathcal{U}$

(r) $\{\{3, 1\}\} \subset \mathcal{U}$

(s) $\{\{b, c\}, b, c\} \subset \mathcal{U}$

(t) $\{\ \} \subset \mathcal{U}$

Question 4.4 Find all the subsets of the following sets. Which of these subsets are proper subsets?

(a) $\{a\}$

(b) $\{5\}$

(c) $\{v\}$

(d) $\{a, b\}$

(e) $\{5, 7\}$

(f) $\{5, e\}$

(g) $\{2, 4, 6\}$

(h) $\{x, y, z\}$

(i) $\{2, b, c\}$

Question 4.5 Find all the subsets of the following sets. Which of these subsets are proper subsets?

(a) $\left\{\{a, b, c\}\right\}$

(b) $\left\{\{x, y\}\right\}$

(c) $\left\{\{1, 2, 3\}\right\}$

(d) $\left\{r, \{s, t\}\right\}$

(e) $\left\{\{x\}, \{y, z\}\right\}$

(f) $\left\{\{2, 4\}, 6\right\}$

(g) $\left\{2, \{4\}, 6\right\}$

(h) $\left\{\{2\}, \{4\}, \{6\}\right\}$

(i) $\left\{u, v, \{w\}\right\}$

Question 4.6 Let $\mathcal{U} = \{0, 1, 2, 3, 4, 5, 6, 7, 8, 9\}$, $A = \{2, 3, 4, 5, 6\}$, and $B = \{4, 5, 6, 7, 8\}$. Find the following sets.

(a) $A \cup B$

(b) $A \cap B$

(c) A^c

(d) B^c

(e) $(A \cup B)^c$

(f) $(A \cap B)^c$

(g) $A^c \cup B^c$

(h) $A^c \cap B^c$

(i) $A - B$

(j) $B - A$

Question 4.7 Let $\mathcal{U} = \{1, 2, 3, 4, 5, a, b, c, d, e\}$, $A = \{1, 2, a, b, c, d\}$, and $B = \{2, 3, 4, c, d, e\}$. Find the following sets.

(a) $A \cup B$

(b) $A \cap B$

(c) A^c

(d) B^c

(e) $(A \cup B)^c$

(f) $(A \cap B)^c$

(g) $A^c \cup B^c$

(h) $A^c \cap B^c$

(i) $A - B$

(j) $B - A$

Question 4.8 Let $\mathcal{U} = \{o, p, q, r, s, t, u, v, w, x, y, z\}$, $A = \{o, p, s, t, u, y\}$, and $B = \{p, r, s, t, u, v\}$. Find the following sets.

(a) $A \cup B$

(b) $A \cap B$

(c) A^c

(d) B^c

(e) $(A \cup B)^c$

(f) $(A \cap B)^c$

(g) $A^c \cup B^c$

(h) $A^c \cap B^c$

(i) $A - B$

(j) $B - A$

Question 4.9 Write the following sets in enumerated form. We may assume x is from the set $\mathbb{Z}$.

(a) $\{x | x^2 = 4\}$

(b) $\{x | (x+1)(x-1) = 0\}$

(c) $\{x | (x-5)(x+4) = 0\}$

(d) $\{x | x^3 = 27\}$

(e) $\{x | x^2 = 25\}$

(f) $\{x | (2x-8)(5x+15) = 0\}$

(g) $\{x | x^4 = 16\}$

(h) $\{x | (x+3)(x-2) = 0\}$

(i) $\{x | x = -x\}$

Question 4.10 *Write the following sets in enumerated form. We may assume x is from the set $\mathbb{R}$.*

(a) $\{x \mid x^3 = -1\}$.

(b) $\{x \mid (2x - 3)(4x + 7) = 0\}$

(c) $\{x \mid (2x + 5)(3x + 4) = 0\}$

(d) $\{x \mid (x^2 - 4)(x - 3) = 0\}$

(e) $\{x \mid (5 - x)(7 + 3x) = 0\}$

(f) $\{x \mid x^3 = 1\}$

(g) $\{x \mid x = 2x\}$

(h) $\{x \mid (3x - 4)(2 - 3x) = 0\}$

(i) $\{x \mid (x + 5)(x^2 - 9) = 0\}$

Question 4.11 *Given the sets*

$$A = \left\{ n \in \mathbb{N} \mid n > 100 \right\},$$

$$B = \left\{ n \in \mathbb{N} \mid n \leq 300 \right\},$$

$$E = \left\{ n \in \mathbb{N} \mid n \text{ is even} \right\},$$

$$O = \left\{ n \in \mathbb{N} \mid n \text{ is odd} \right\}$$

find the cardinality of the following sets.

(a) A^c

(b) B^c

(c) $A^c \cap B^c$

(d) O^c

(e) $O \cap E$

(f) E^c

(g) $A \cup B$

(h) $B \cap E$

(i) $A^c \cap O$

(j) $O \cap E^c$

(k) $A \cap O^c$

(l) $B \cup E$

Question 4.12 *If $A = \{1, 2, 3, 4, 5, 6\}$ how many subsets does A have? How many proper subsets does A have?*

Question 4.13 *If $B = \{a, b, c, d, w, x, y, z\}$ how many subsets does B have? How many proper subsets does B have?*

Question 4.14 *If $C = \{1, 3, 5, 7, 9, r, s, t, u, v\}$ what is the cardinality of the power set of of C?*

Question 4.15 *If $D = \{1, 2, 3, 4, 5, \ldots, 20\}$ what is the cardinality of the power set of of D?*

Question 4.16 *If $A = \{a, b, c\}$ and $B = \{1, 2, 3\}$ find the Cartesian product $A \times B$.*

Question 4.17 *If $S = \{e, f, g, h\}$ and $T = \{9, 8, 7, 6\}$ find the Cartesian product $S \times T$.*

Question 4.18 *If $X = \{0, 1, 2, 3, 4, 5\}$ and $Y = \{t, f\}$ find the Cartesian product $X \times Y$.*

Question 4.19 *If $X_1 = \{0, 1\}$, $X_2 = \{0, 1\}$, $X_3 = \{0, 1\}$, and $X_4 = \{0, 1\}$ find the Cartesian product $X_1 \times X_2 \times X_3 \times X_4$.*

Question 4.20 *Using the Venn diagram find the sets A, B, A∪B, A∩B, A^c, B^c, (A∪B)^c, (A∩B)^c, A^c∪B, and A∪B^c, A − B, B − A.*

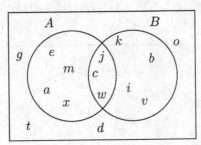

Question 4.21 *Using the Venn diagram find the sets A, B, A∪B, A∩B, A^c, B^c, (A∪B)^c, (A∩B)^c, A^c∪B, and A∪B^c, A − B, B − A.*

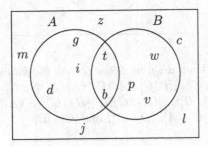

Question 4.22 *Using the Venn diagram below to find the following: A, B, C, A∩B, A∩C, B∩C, A∪B, A∪C, B∪C, A∩B∩C, A∪B∪C, A^c, B^c, C^c, (A∩B)^c, (A∩C)^c, (B∩C)^c, (A∪B)^c, (A∪C)^c, (B∪C)^c, (A∩B∩C)^c, (A∪B∪C)^c, A − B, B − A, A − C, C − A, B − C, and C − B.*

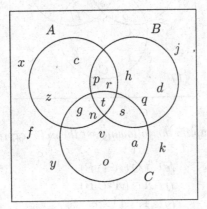

Question 4.23 *Using the Venn diagram below to find the following:* $A, B, C, A \cap B, A \cap C,$ $B \cap C, A \cup B, A \cup C, B \cup C, A \cap B \cap C, A \cup B \cup C, A^c, B^c, C^c, (A \cap B)^c, (A \cap C)^c,$ $(B \cap C)^c, (B \cup C)^c, (A \cup C)^c, (B \cup C)^c, (A \cap B \cap C)^c, (A \cup B \cup C)^c, A - B, B - A, A - C,$ $C - A, B - C, and \ C - B$

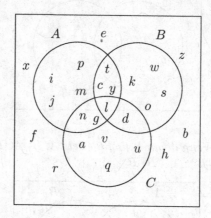

Question 4.24 *Using the Venn diagram below to find the following:* $|A|, |B|, |C|, |A \cap B|,$ $|A \cap C|, |B \cap C|, |A \cup B|, |A \cup C|, |B \cup C|, |A \cap B \cap C|, |A \cup B \cup C|, |A^c|, |B^c|, |C^c|, |(A \cap B)^c|,$ $|(A \cap C)^c|, |(B \cap C)^c|, |(A \cup B)^c|, |(A \cup C)^c|, |(B \cup C)^c|, |(A \cap B \cap C)^c|, |(A \cup B \cup C)^c|,$ $|A - B|, |B - A|, |A - C|, |C - A|, |B - C|, and \ |C - B|$

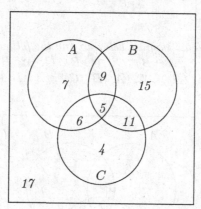

Question 4.25 *(Easy) Simplify the following set theoretic expressions using the laws of set theory.*

(a) $A \cup A$

(b) $A \cup A^c$

(c) $(A \cup A^c)^c$

(d) $A \cup (A \cap B)$

(e) $A \cup (A^c \cap B)$

(f) $A \cap (A^c \cup B)$

(g) $A \cap A$

(h) $A \cap (A \cup B)$

(i) $A \cap (A \cup B \cup C)$

(j) $(A \cap B) \cup (A \cap B^c)$

(k) $(A^c \cup A^c)^c$

(l) $(A \cap B) \cup (A^c \cap B)$

Question 4.26 *(Moderate) Simplify the following set theoretic expressions using the laws of set theory.*

(a) $(A^c \cup B^c) \cap (A^c \cup B)$

(b) $B \cup (B \cap B^c)$

(c) $A^c \cup (B \cap A^c)$

(d) $(A \cup B^c) \cap (A \cup B)$

(e) $C \cup [C \cup (C \cap A)]$

(f) $A \cap [A \cup (A \cap B)]$

(g) $C \cup (C \cap A^c \cap B \cap D)$

(h) $C^c \cap (C \cap A \cap B \cap D)^c$

Question 4.27 *(Difficult) Simplify the following set theoretic expressions using the laws of set theory.*

(a) $B \cap [A \cup (A^c \cap B)]$

(b) $[(A \cap B^c) \cup A^c]^c$

(c) $(A^c \cap B) \cup [A \cap (A \cup B)]$

(d) $[A \cup (A \cap B)]^c \cap B$

(e) $(A \cap C^c) \cup (A^c \cap B) \cup (B \cap C)^c$

(f) $(A \cap B)^c \cup (B \cap C)$

(g) $[A \cup (B \cap C)] \cap (A^c \cup C)$

(h) $(A \cap C^c) \cup (A^c \cap B) \cup (A \cap C)^c$

(i) $(A \cap C) \cup (A^c \cap B) \cup (C \cap B)$

(j) $A^c \cup B^c \cup (A \cap B \cap C^c)$

(k) $(A \cup C) \cap (A^c \cup B) \cap (C \cup B)$

(l) $(A \cap B)^c \cap (A^c \cup B) \cap (B^c \cup B)$

Question 4.28 *Given the sets $A = \{1, 2, 3, 4\}$ and $B = \{1, 2, 3, 4\}$ write out the sets $R \subset A \times B$ that correspond to the following relations:*

(a) *is equal to*

(b) *is less than*

(c) *is greater than*

(d) *is less than or equal*

(e) *is greater than or equal*

(f) *is not equal to*

Question 4.29 *Given the sets $A = \{0, 2, 3, 5, 7\}$ and $B = \{1, 2, 4, 5, 6\}$ write out the sets $R \subset A \times B$ that correspond to the following relations:*

(a) *is equal to*

(b) *is less than*

(c) *is greater than*

(d) *is less than or equal*

(e) *is greater than or equal*

(f) *is not equal to*

Question 4.30 *The below directed graph represents a relation. Find the domain and range of the relation. Then write the relation as a subset of the Cartesian product of the domain and range.*

Question 4.31 *Find the stated relations on the set $\{1, 2, 3\}$. Then determine if they are reflexive, irreflexive, symmetric, antisymmetric, or transitive.*

(a) *the relation $\leq$*

(b) *the relation $<$*

(c) *the relation $=$*

(d) *the relation $\neq$*

(e) *the relation $>$*

(f) *the relation $\geq$*

Question 4.32 *Show the relation "is greater than or equal to" is a partial ordering on $\mathbb{N}$.*

Question 4.33 *Let $A = \{1, 2, 3, 4, 5, 6\}$ with the relation given by*

$$R = \{(1, 1), (1, 5), (5, 1), (5, 5), (2, 2), (2, 4), (2, 6)$$
$$(4, 2), (4, 4), (4, 6), (6, 2), (6, 4), (6, 6), (3, 3)\} \subset A \times A.$$

Show this relation is an equivalence relation. Then find the equivalence classes that partition the set A.

Question 4.34 *Show the relation "is equal to" is both a partial ordering and an equivalence relation on $\mathbb{N}$.*

CHAPTER 5

Boolean Algebra

There are two major reasons to study Boolean algebra in computer science. First, Boolean algebra unifies both logic and set theory in a very nice way. Second, Boolean algebra plays a very important role in circuit design in computer science. It is very helpful for computer science majors to have some idea of how circuits are designed and work. Though this topic is beyond the scope of this book, appendix A shows how the the ideas from Boolean algebra can be used to design a circuit that adds two eight-digit binary numbers.

5.1 DEFINITION OF BOOLEAN ALGEBRA

We will begin by seeing how Boolean algebra unifies logic and set theory. In logic, even though we had propositions and expressions, what we really were interested in is the truth-values. Remember, in the truth tables we connected truth-values (T or F) with connectives like $\wedge$, $\vee$, and $\neg$ to get a new truth-value. We then found a variety of equivalence relations we called laws. In summary, we had

- a set $\{T, F\}$,
- three operations $\wedge$, $\vee$, and $\neg$ defined on elements of the set $\{T, F\}$,
- a list of laws the operations satisfied.

In set theory we start out with a universal set $\mathcal{U}$ but then define the power set of $\mathcal{U}$, $\mathcal{P}(\mathcal{U})$. The power set $\mathcal{P}(\mathcal{U})$ is the set of all subsets of $\mathcal{U}$. We then operate on these elements of $\mathcal{P}(\mathcal{U})$ using $\cap$, $\cup$, and ^{c}. Again, we found a variety of equalities we called laws. In summary, we had

- a set $\mathcal{P}(\mathcal{U})$,
- three operations $\cap$, $\cup$, and ^{c} defined on elements of the set $\mathcal{P}(\mathcal{U})$,
- a list of laws the operations satisfy.

So, we can see that in a lot of ways logic and set theory are very similar. And when you look at the list of laws in logic and in set theory you see that even these laws are very similar. In fact, if we exchanged the $\wedge$ with $\cap$, the $\vee$ with $\cup$, the $\neg$ with ^{c}, the F with $\varnothing$, and the T with $\mathcal{U}$ the laws are exactly the same. This is because they are both examples of what is called a Boolean algebra.

Instead of studying logic or set theory as we have done up till now, we can study the laws themselves. When we study the laws themselves we are studying both logic and set theory and many other things at the same time. Whatever we prove using Boolean algebra must also be true for logic or set theory. Now we are ready for the definition.

Definition 5.1 A **Boolean algebra** is a set B that has two binary operations, $+$ called **addition**, $\times$ called **multiplication**, and a unary operation $'$ called **complement**. The set B contains two special elements written as 0 and 1 such that the following **axioms** are true for all $x, y, z \in B$:

Name	Axioms of Boolean Algebra	
Commutative axiom	(a) $x \times y = y \times x$	(b) $x + y = y + x$
Associative axiom	(a) $x \times (y \times z) = (x \times y) \times z$	(b) $x + (y + z) = (x + y) + z$
Distributive axiom	(a) $x \times (y+z) = (x \times y) + (x \times z)$	(b) $x + (y \times z) = (x+y) \times (x+z)$
Identity axiom	(a) $x \times 1 = x$	(b) $x + 0 = x$
Inverse axiom	(a) $x \times x' = 0$	(b) $x + x' = 1$

Notice the definition does not say what the elements of B are, only that there are two special elements called 0 and 1. There might be other elements in the Boolean algebra set or there might not be. What the elements of B are depends on how we **interpret** the set B. Also, our Boolean operations **addition**, **multiplication**, and **complement** may be interpreted differently as well, depending on what the elements of B actually are. These operations are NOT the addition and multiplication you know from grade school. We used the same words but they mean something else.

We also use the word **axiom** instead of law because these ten equations are part of the definition of Boolean algebra. Why did we use only these five axioms in the definition instead of more? We are missing the double complement laws, the idempotent laws, DeMorgan's laws, the annihilation laws, and the absorption laws. All these other laws can be proved from the axioms in the definition. We list the rest of the laws:

Name	Laws of Boolean Algebra	
Double negation law	$x'' = x$	
Idempotent laws	(a) $x \times x = x$	(b) $x + x = x$
DeMorgan's laws	(a) $(x \times y)' = x' + y'$	(b) $(x + y)' = x' \times y'$
Annihilation laws	(a) $x \times 0 = 0$	(b) $x + 1 = 1$
Absorption laws	(a) $x \times (x + y) = x$	(b) $x + (x \times y) = x$
Complement laws	(a) $1' = 0$	(b) $0' = 1$

Now we will see how to prove some of the Boolean algebra laws from the axioms given in the definition of Boolean algebra.

Example 5.1

Prove the idempotent law for addition, $x + x = x$, using the Boolean algebra axioms.

$$
\begin{aligned}
x + x &= (x + x) \times 1 & \text{second identity axiom} \\
&= (x + x) \times (x + x') & \text{first inverse axiom} \\
&= x + (x \times x') & \text{first distributive axiom} \\
&= x + 0 & \text{second inverse axiom} \\
&= x & \text{first identity axiom}
\end{aligned}
$$

It takes a little while to get used to seeing something like $x + x = x$. Just remember, the $+$ is not the addition you are used to from algebra.

If we have an expression in Boolean algebra we find the **dual** of the expression by replacing every $+$ with $\times$, every $\times$ with $+$, every 0 with 1, and every 1 with 0.

Example 5.2

Finding the dual of an expression

- The dual of $x \times (x \times x')'$ is $x + (x + x')'$.
- The dual of $(x + y) \times (x \times y')$ is $(x \times y) + (x + y')$.
- The dual of $x + x = x$ is $x \times x = x$.

The reason the dual is important is the following theorem, which we will not prove.

Theorem 5.1 *Duality Principle*: *The dual of anything proved to be true in Boolean algebra is also true.*

Example 5.3

Show the idempotent law for multiplication, $x \times x = x$.

From the last two examples we already know that $x + x = x$ is true and the dual of this statement is $x \times x = x$. Thus by the duality principle $x \times x = x$ is also true.

Example 5.4

Prove the annihilation law using the Boolean algebra axioms.

$$
\begin{aligned}
x \times 0 &= (x \times 0) + 0 & &\text{first identity axiom} \\
&= (x \times 0) + (x \times x') & &\text{second inverse axiom} \\
&= x \times (0 + x') & &\text{second distributive axiom} \\
&= x \times (x' + 0) & &\text{first commutative axiom} \\
&= x \times x' & &\text{first identity axiom} \\
&= 0 & &\text{second inverse axiom}
\end{aligned}
$$

Thus we have shown $x \times 0 = 0$. The dual of this is $x + 1 = 1$, which is also true by the duality principle.

Example 5.5

Prove the absorption law.

$$
\begin{aligned}
x \times (x + y) &= (x + 0) \times (x + y) & &\text{first identity axiom} \\
&= x + (0 \times y) & &\text{first distributive axiom} \\
&= x + (y \times 0) & &\text{second commutitive axiom} \\
&= x + 0 & &\text{annihilation law (already proved)} \\
&= x & &\text{first identity axiom}
\end{aligned}
$$

Thus we have shown $x \times (x + y) = x$. The dual of this is $x + (x \times y) = x$, which is also true by the duality principle.

The other laws are a little more complicated so we will not do those here. However, they can all be proved using the axiom and the laws already proved. These axioms and laws can be used to simplify complicated Boolean algebra expressions.

Example 5.6

Simplify the Boolean expression $(x + x \times y)' \times y$.

$$
\begin{aligned}
(x + x \times y)' \times y &= x' \times (x \times y)' \times y & \text{DeMorgan's law} \\
&= x' \times y \times (x \times y)' & \text{commutitive axiom} \\
&= x' \times y \times (x' + y') & \text{DeMorgan's law} \\
&= (x' \times y \times x') + (x \times y \times y') & \text{distributive axiom} \\
&= (x' \times x' \times y) + (x \times y \times y') & \text{commutitive axiom} \\
&= (x' \times x' \times y) + (x \times 0) & \text{inverse axiom} \\
&= (x' \times y) + (x \times 0) & \text{idempotant law} \\
&= (x' \times y) + 0 & \text{annihilation law} \\
&= x' \times y & \text{identity axiom}
\end{aligned}
$$

Example 5.7

Simplify the Boolean expression $x \times (x \times y')'$.

$$
\begin{aligned}
x \times (x \times y')' &= x \times (x' + y'') & \text{DeMorgan's law} \\
&= x \times (x' + y) & \text{double negation law} \\
&= (x \times x') + (x \times y) & \text{distributive axiom} \\
&= 0 + (x \times y) & \text{inverse axiom} \\
&= x \times y & \text{identity axiom}
\end{aligned}
$$

5.2 LOGIC AND SET THEORY AS BOOLEAN ALGEBRAS

We want to just take a few moments to make sure the connection between logic, set theory, and Boolean algebra is clear. Recall, in logic we had

- A set $\{T, F\}$, the possible truth-values of propositions.[1]
- Three operations $\wedge$, $\vee$, and $\neg$ defined on elements of the set $\{T, F\}$ (usually denoted as p, q, and r).
- A list of laws the operations satisfied.

---

[1]Another way of thinking about this is that our set is the set of all propositions which can be split into two equivalence classes $T = \{$all true propositions$\}$ and $F = \{$all false propositions$\}$. Thus T is really the equivalence class of all true propositions and F is the equivalence class of all false propositions. Then we could consider p, q, and r to be propositions instead of truth-values.

Name	Laws of Logic	
Commutative laws	(a) $p \wedge q \equiv q \wedge p$	(b) $p \vee q \equiv q \vee p$
Associative laws	(a) $(p \wedge q) \wedge r \equiv p \wedge (q \wedge r)$	(b) $(p \vee q) \vee r \equiv p \vee (q \vee r)$
Distributive laws	(a) $p \wedge (q \vee r) \equiv (p \wedge q) \vee (p \wedge r)$	(b) $p \vee (q \wedge r) \equiv (p \vee q) \wedge (p \vee r)$
Identity laws	(a) $p \wedge T \equiv p$	(b) $p \vee F \equiv p$
Inverse laws	(a) $p \wedge \neg p \equiv F$	(b) $p \vee \neg p \equiv T$
Double negation law	$\neg\neg p \equiv p$	
Idempotent laws	(a) $p \wedge p \equiv p$	(b) $p \vee p \equiv p$
DeMorgan's laws	(a) $\neg(p \wedge q) \equiv \neg p \vee \neg q$	(b) $\neg(p \vee q) \equiv \neg p \wedge \neg q$
Annihilation laws	(a) $p \wedge F \equiv F$	(b) $p \vee T \equiv T$
Absorption laws	(a) $p \wedge (p \vee q) \equiv p$	(b) $p \vee (p \wedge q) \equiv p$
Negation laws	(a) $\neg T = F$	(b) $\neg F = T$

In set theory we had

- A set $\mathcal{P}(\mathcal{U})$, the power set of the set $\mathcal{U}$.
- Three operations $\cap$, $\cup$, and ^{c} defined on elements of the set $\mathcal{P}(\mathcal{U})$ (usually denoted as A, B, and C).
- A list of laws the operations satisfy.

Name	Laws of Set Theory	
Commutative laws	(a) $A \cap B = B \cap A$	(b) $A \cup B = B \cup A$
Associative laws	(a) $(A \cap B) \cap C = A \cap (B \cap C)$	(b) $(A \cup B) \cup C = A \cup (B \cup C)$
Distributive laws	(a) $A \cap (B \cup C) = (A \cap B) \cup (A \cap C)$	(b) $A \cup (B \cap C) = (A \cup B) \cap (A \cup C)$
Identity laws	(a) $A \cap \mathcal{U} = A$	(b) $A \cup \varnothing = A$
Inverse laws	(a) $A \cap A^c = \varnothing$	(b) $A \cup A^c = \mathcal{U}$
Double complement law	$(A^c)^c = A$	
Idempotent laws	(a) $A \cap A = A$	(b) $A \cup A = A$
DeMorgan's laws	(a) $(A \cap B)^c = A^c \cup B^c$	(b) $(A \cup B)^c = A^c \cap B^c$
Annihilation laws	(a) $A \cap \varnothing = \varnothing$	(b) $A \cup \mathcal{U} = \mathcal{U}$
Absorption laws	(a) $A \cap (A \cup B) = A$	(b) $A \cup (A \cap B) = A$
Complement laws	(a) $\mathcal{U}^c = \varnothing$	(b) $\varnothing^c = \mathcal{U}$

And in Boolean algebra we had

- A set $\mathcal{B}$ which included two special elements we called 1 and 0.
- Three operations $\times$, $+$, and $'$ defined on elements of the set $\mathcal{B}$ (usually denoted as x, y, and z).
- A list of axioms and laws the operations satisfy.

Name	Axioms of Boolean Algebra	
Commutative axiom	(a) $x \times y = y \times x$	(b) $x + y = y + x$
Associative axiom	(a) $x \times (y \times z) = (x \times y) \times z$	(b) $x + (y + z) = (x + y) + z$
Distributive axiom	(a) $x \times (y+z) = (x \times y) + (x \times z)$	(b) $x + (y \times z) = (x+y) \times (x+z)$
Identity axiom	(a) $x \times 1 = x$	(b) $x + 0 = x$
Inverse axiom	(a) $x \times x' = 0$	(b) $x + x' = 1$
Name	**Laws of Boolean Algebra**	
Double negation law	$x'' = x$	
Idempotent laws	(a) $x \times x = x$	(b) $x + x = x$
DeMorgan's laws	(a) $(x \times y)' = x' + y'$	(b) $(x + y)' = x' \times y'$
Annihilation laws	(a) $x \times 0 = 0$	(b) $x + 1 = 1$
Absorption laws	(a) $x \times (x + y) = x$	(b) $x + (x \times y) = x$
Complement laws	(a) $1' = 0$	(b) $0' = 1$

In mathematics we would say that Boolean algebra is the theoretical framework that encompass both logic and set theory. This is just a fancy way of saying both logic and set theory are concrete examples of Boolean algebras. By comparing the laws in each case we can see that the following symbols are equivalent.

Boolean Algebra	Logic	Set Theory
$\times$ (multiplication)	$\wedge$ (and)	$\cap$ (intersection)
$+$ (addition)	$\vee$ (or)	$\cup$ (union)
$'$ (complement)	$\neg$ (negation)	^{c} (complement)
1	T	$\mathcal{U}$
0	F	$\varnothing$

As long as we remember the equivalence between these symbols we only have to learn one set of laws, not three.

As computer scientists we are especially interested in one concrete example of a Boolean algebra where

$$\mathcal{B} = \Big\{ \text{no electrical current}, \text{electrical current} \Big\}.$$

It is this Boolean algebra that is used by computer engineers to design the circuits that run all computers. But of course, we usually make the following identification,

$$\text{no electrical current} = 0,$$

$$\text{electrical current} = 1,$$

which means we usually write our Boolean set as $\mathcal{B} = \{0, 1\}$.[2] Recall, the definition said that a Boolean algebra $\mathcal{B}$ must contain two special elements called 0 and 1. The simplest possible Boolean algebra has only these two special elements and no other elements. So, the simplest Boolean algebra is given by $\mathcal{B} = \{0, 1\}$. This is exactly the Boolean algebra that we computer scientists are most interested in.

We will use the laws and axioms to see how Boolean algebra addition and multiplication work in $\mathcal{B} = \{0, 1\}$. The first identity axiom says that $x \times 1 = x$. Since x can either be 1 or 0 we have

$$1 \times 1 = 1,$$

$$0 \times 1 = 0.$$

---

[2]This is a simplification. Though it is nice to think of 0 meaning "no electrical current" there is actually a very low level of residual current.

The second identity axiom says $x + 0 = x$ so we have

$$1 + 0 = 1,$$
$$0 + 0 = 0.$$

The first commutative axiom says $x \times y = y \times x$ so using this and the fact that $0 \times 1 = 0$ we know

$$0 \times 1 = 1 \times 0$$
$$\Rightarrow 1 \times 0 = 0.$$

Similarly, the second commutative axiom says $x + y = y + x$ so using this and the fact that $1 + 0 = 1$ we know

$$1 + 0 = 0 + 1$$
$$\Rightarrow 0 + 1 = 1.$$

Finally, we use the first annihilation law $x \times 0 = 0$ with $x = 0$ to get

$$0 \times 0 = 0$$

and the second annihilation law $x + 1 = 1$ with $x = 1$ to get

$$1 + 1 = 1.$$

Putting all this together we have the following multiplication and addition tables for Boolean algebra,

$$0 \times 0 = 0,$$
$$1 \times 0 = 0,$$
$$0 \times 1 = 0,$$
$$1 \times 1 = 1,$$

and

$$0 + 0 = 0,$$
$$1 + 0 = 1,$$
$$0 + 1 = 1,$$
$$1 + 1 = 1.$$

Everything is as you would expect from grade school math, except the funny $1 + 1 = 1$. This is what makes Boolean algebra different from grade school math. But this addition rule $1 + 1 = 1$ makes a little more sense when we think of it in terms of electricity, since

$$\text{electrical current} + \text{electrical current} = \text{electrical current}.$$

Here the multiplication and addition tables are written in a slightly different way:

$\times$	0	1
0	0	0
1	0	1

and

$+$	0	1
0	0	1
1	1	1

5.3   DIGITAL CIRCUITS

Now we turn to the use of Boolean algebra for designing computer **digital circuits**. Digital circuits operate with electricity. Either there is an electrical current or there isn't. So there are only two possibilities for a signal in a computer, no current or current. We call those possibilities 0 or 1.

At the most basic level digital circuits combine electrical signals to produce a new electrical signal. The circuit does this at a **gate**. We will not describe how this is physically done in the computer, that is the job of electrical engineers. We will only say that circuits are made of gates that combine signals to make a new signal. We can model these gates with a Boolean algebra $B = \{0, 1\}$. It turns out that this Boolean algebra is just like the Boolean algebra for logic. We let $\times$ represent **and**, we let $+$ represent **or**, and we let $'$ represent **not**.

First, it is always possible for a circuit to "flip" a signal. That is, it can change a 0 to a 1 or a 1 to a 0. That is done with a **not gate**. Just like in logic we can build a table that shows how the not gate changes a signal. Even though these tables do not contain trues or falses we will still call them truth tables because they are so similar.

x	x'
1	0
0	1

It turns out to be very useful to use pictures to help computer scientists design circuits using gates. Here is the picture that is used to indicate a **not gate**. It is a triangle with a tiny circle attached to the right point. As it is drawn the input is signal x and the output is signal x'. We generally "read" circuit diagrams going from left to right just like we read English.

Next we will consider gates with two input signals x and y. We want to consider the different ways those two input signals can be combined. Of course x can be either 1 or 0 and y can also be either 1 or 0. So there are $2^2 = 4$ possible combinations of inputs. Each gate needs to give a unique output for each of these four inputs.

x	y	gate output
1	1	0 or 1
1	0	0 or 1
0	1	0 or 1
0	0	0 or 1

There are a few combinations of outputs we are not interested in. We are not interested in a gate that gives 0 no matter what the inputs are. Also, we are not interested in a gate that gives 1 no matter what the inputs are. Finally, we do not want order to matter. That means we want the same output for when $x = 0$ and $y = 1$ as for when $x = 1$ and $y = 0$. With these restrictions there are only six possible gates.

The first two-input gate we will look at is the **and gate**. The and gate works just like the **and** connector in logic. However, we use $\times$ to indicate **and** in Boolean algebra. We can also write $x \times y$ as xy just as one would in algebra. The truth table for the **and gate** is given below.

x	y	$x \times y$
1	1	1
1	0	0
0	1	0
0	0	0

Here is the picture we use to show an **and gate**. There are two inputs, x and y that enter the gate on the left. The output from the gate is shown on the right and is given by $x \times y$.

The next two-input gate we will look at is the **or gate**. The **or gate** works just like the **or** connector in logic. However, we use $+$ to indicate **or** in Boolean algebra. The truth table for the **or gate** is given below.

x	y	$x + y$
1	1	1
1	0	1
0	1	1
0	0	0

Here is the picture we use to show an **or gate**. There are two inputs, x and y that enter the gate on the left. The output from the gate is shown on the right and is given by $x + y$.

$$x + y$$

Now we will look at the **nand gate**. The **nand gate** is the **not** of the **and gate**,

$$x \text{ nand } y = (x \times y)'.$$

We will simply use the term **nand** to mean the **nand** operation. The truth table for the **nand gate** is given below.

x	y	$x \text{ nand } y$
1	1	0
1	0	1
0	1	1
0	0	1

Here is the picture we use to show an **nand gate**. There are two inputs, x and y that enter the gate on the left. The output from the gate is shown on the right and is given by $x \text{ nand } y$. Notice that the picture for an **nand gate** looks like an **and gate**, only has a tiny circle on the right where the output comes out.

$$x \text{ nand } y$$

Next we look at the **nor gate**. The **nor gate** is the **not** of the **or gate**,

$$x \ \textbf{nor} \ y = (x + y)'.$$

We will simply use the term **nor** to mean the **nor** operation. The truth table for the **nor gate** is given.

x	y	x nor y
1	1	0
1	0	0
0	1	0
0	0	1

Here is the picture we use to show a **nor gate**. There are two inputs, x and y that enter the gate on the left. The output from the gate is shown on the right and is given by x **nor** y. Again, notice that the picture for a **nor gate** looks like an **or gate**, only has a tiny circle on the right where the output comes out.

The next gate we look at is called the **exclusive-or gate**, or **xor gate**. In an example from the logic chapter we said the expression p **xor** q is true when only one of the propositions p or q is true. In other words, p **xor** q is true when one, or the other, of the propositions are true but not both. That means that if both p and q are true then p **xor** q is false. Similarly, x **xor** y is equal to 1 only when either x or y is 1 but not both. We can also write this as

$$x \ \textbf{xor} \ y = x \times y' + x' \times y.$$

Sometimes $\oplus$ is also used for **xor**.

x	y	x xor y
1	1	0
1	0	1
0	1	1
0	0	0

Here is the picture we use to show a **xor gate**. There are two inputs, x and y that enter the gate on the left. The output from the gate is shown on the right and is given by x **xor** y. Again, notice that the picture for an **xor gate** looks like an **or gate** only with an extra line on the left where the input go into the gate.

Finally, we look at the **xnor gate**. The **xnor gate** is the **not** of the **xor gate**. That is,

$$x \ \textbf{xnor} \ y = (x \times y' + x' \times y)' = (x \times y) + (x' \times y').$$

We will simply use the term **xnor** to mean the **xnor** operation. The truth table for the **xnor gate** is given.

x	y	x **xnor** y
1	1	1
1	0	0
0	1	0
0	0	1

Here is the picture we use to show an **xnor gate**. There are two inputs, x and y that enter the gate on the left. The output from the gate is shown on the right and is given by x **xnor** y. Again, notice that the picture for an **xnor gate** looks like an **xor gate** only with a tiny circle on the right where the output comes out.

$$x \text{ xnor } y$$

We provide all six of these Boolean algebra operations, or gates, in a single table. As we said before, we were not interested in gates that gave a 0 or a 1 no matter what the inputs were. And we are only interested in gates that give an output that are the same when $x = 0$ and $y = 1$ as when $x = 1$ and $y = 0$.

x	y	$x \times y$	$x + y$	x **nand** y	x **nor** y	x **xor** y	x **xnor** y
1	1	1	1	0	0	0	1
1	0	0	1	1	0	1	0
0	1	0	1	1	0	1	0
0	0	0	0	1	1	0	1

Drawing circuits is almost exactly like drawing the expression trees for expressions in logic. It is these "pictures" that actually are the basis for the circuits that are etched into computer microprocessors, the "chips" that computers run on. We will draw a circuit for evaluating the expression $x \times y' + y$. We begin by drawing an expression tree similar to the expression trees from logic.

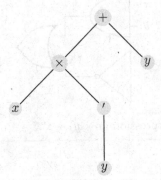

There are a couple differences between circuit diagrams and expression trees. First, expression trees are usually read from the bottom to the top. Circuit diagrams are usually read from the left to the right. Second, expression trees often have the same variable listed more than once. In the above expression tree the variable y is shown twice, once for each time it appears in the expression $x \times y' + y$. But circuit diagrams generally have each input signal shown only once. Below we rotate the expression tree so it can be read left to right and modify it a little so the y variable only appears once. (Notice, it no longer looks like a tree after we modify it.)

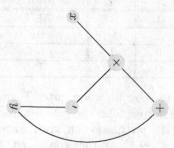

This modified expression tree can be used as a basis for drawing the circuit diagram. The vertices with the variables become the input signals. Each of the other vertices of the expression tree is replaced by the correct gate. The $'$ is replaced with the **not gate**, the $\times$ is replaced with the **and gate**, and the $+$ is replaced with the **or gate**.

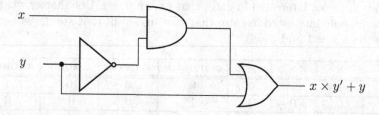

Finding the expression tree can be helpful for drawing circuit diagrams but is not necessary once you get good at it.

Example 5.8

What is the expression that this circuit represents?

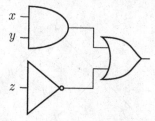

This circuit represents the expression $(x \times y) + z'$.

Example 5.9

Draw the circuit for the expression $(x' + y') \times (x + y') \times (y + z)$.

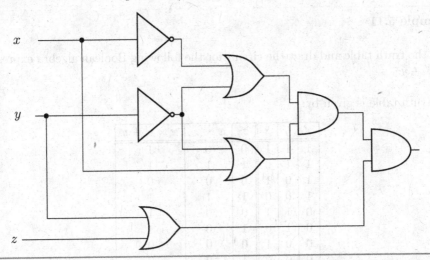

Example 5.10

The circuit drawn in the last example is quite complicated. Use the laws of Boolean algebra to simplify the expression $(x' + y') \times (x + y') \times (y + z)$ and then draw the circuit for the simplified expression.

$$
\begin{aligned}
(x' + y')(x + y')(y + z) &= (x'x + x'y' + y'x + y'y')(y + z) \\
&= [0 + y'(x' + x) + y'](y + z) \\
&= (y'1 + y')(y + z) \\
&= (y' + y')(y + z) \\
&= y'(y + z) \\
&= y'y + y'z \\
&= y'z
\end{aligned}
$$

Which can be drawn as

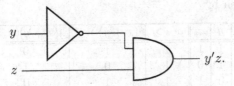

$y'z$.

This circuit is easier and cheaper to build and faster than the circuit shown in the last example. This shows why simplifying Boolean algebra expressions is so important in computer science.

5.4   SUMS-OF-PRODUCTS AND PRODUCTS-OF-SUMS

Just like in logic we can find the truth table for Boolean algebra expressions. But instead of using T or F like in logic we use 0 and 1 in Boolean algebra.

Example 5.11

Find the truth table and draw the circuit for the following Boolean algebra expression $x \times z' + y$.

The truth table is given by:

x	y	z	z'	$x \times z'$	$x \times z' + y$
1	1	1	0	0	1
1	1	0	1	1	1
1	0	1	0	0	0
1	0	0	1	1	1
0	1	1	0	0	1
0	1	0	1	0	1
0	0	1	0	0	0
0	0	0	1	0	0

The circuit diagram is given by:

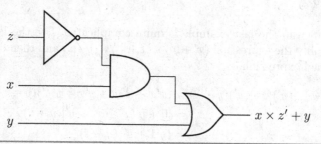

Example 5.12

Find the truth table and draw the circuit for the following Boolean algebra expression $[(x + y) \times z]'$.

The truth table is given by:

x	y	z	$x + y$	$(x + y) \times z$	$[(x + y) \times z]'$
1	1	1	1	1	0
1	1	0	1	0	1
1	0	1	1	1	0
1	0	0	1	0	1
0	1	1	1	1	0
0	1	0	1	0	1
0	0	1	0	0	1
0	0	0	0	0	1

The circuit diagram is given by:

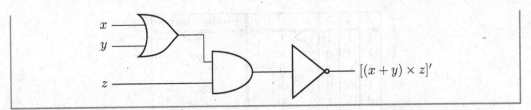

Suppose you were given two expressions, $(x'x + x'y' + y'x + y'y')(y + z)$ and $y'(y + z)$. Are these expressions equal or not? That is, do these expressions represent the same thing or not? How would you find out? One thing you could do is write out a truth table for each and see if they are equivalent. In this case they are equivalent. In a previous example we simplified the expression $(x' + y')(x + y')(y + z)$,

$$
\begin{aligned}
(x' + y')(x + y')(y + z) &= (x'x + x'y' + y'x + y'y')(y + z) \\
&= [0 + y'(x' + x) + y'](y + z) \\
&= (y'1 + y')(y + z) \\
&= (y' + y')(y + z) \\
&= y'(y + z) \\
&= y'y + y'z \\
&= y'z.
\end{aligned}
$$

The two expressions $(x'x + x'y' + y'x + y'y')(y + z)$ and $y'(y + z)$ are simply two different lines in this above simplification. In fact, each of the above lines is a different way of writing down the same thing.

Circuit designers need an easy way to know if different expressions really are the same thing or not. Therefore circuit designers often write expressions in one of two standardized, or canonical, ways. Every Boolean expression can be written in a sum-of-products form or in a product-of-sums form. This makes it easy to see when expressions are the same or not.

Additionally, sometimes you want to have a function that behaves in a certain way. Suppose you know the truth table for what you want the function to do but do not know a function that behaves that way. You want to find an expression for a function that behaves that way. The sum-of-products and product-of-sum expressions can be written down using a truth table.

We will first consider the **sum-of-products** form. The first step in writing an expression in the sum-of-products form is to find the truth table. We use the truth table for $x \times z' + y$ from the above example. When there is a 1 in the final column we consider the values in the row.

- If x value is 1 use x in the product term, if x value is 0 use x' in the product term.
- If y value is 1 use y in the product term, if y value is 0 use y' in the product term.
- If z value is 1 use z in the product term, if z value is 0 use z' in the product term.

We then add up, or sum, all the product terms we get. This is the sum-of-products form for the expression.

x	y	z	$x \times z' + y$	product terms
1	1	1	1	xyz
1	1	0	1	xyz'
1	0	1	0	
1	0	0	1	$xy'z'$
0	1	1	1	$x'yz$
0	1	0	1	$x'yz'$
0	0	1	0	
0	0	0	0	

In the first row there is a 1 in the final column so we use this row. In this row the x value is 1, the y value is 1, and the z value is 1 so we use x, y, and z in the product term to get $x \times y \times z$. From now on we will write $x \times y \times z$ as xyz. In the second row there is also a 1 in the final column so we use the second row. In this row the x value is 1, the y value is 1, and the z value is 0 so we use x, y, and z' in the product term to get xyz'. In the fourth row there is a 1 in the final column so we use this row. In this row the x value is 1, the y value is 0, and the z value is 0 so we use x, y', and z' in the product term to get $xy'z'$. And so on. We then add up, or sum, all the product terms to get $x \times z' + y = xyz + xyz' + xy'z' + x'yz + x'yz'$.

Example 5.13

Write the sum-of-products form for $[(x + y) \times z]'$.

x	y	z	$[(x + y) \times z]'$	product term
1	1	1	0	
1	1	0	1	xyz'
1	0	1	0	
1	0	0	1	$xy'z'$
0	1	1	0	
0	1	0	1	$x'yz'$
0	0	1	1	$x'y'z$
0	0	0	1	$x'y'z'$

So $[(x + y) \times z]' = xyz' + xy'z' + x'yz' + x'y'z + x'y'z'$.

Next we will consider the **product-of-sum** form. The first step in writing an expression into the product-of-sum form is to find the truth table. Again we use the truth table for $x \times z' + y$ from the above example. When there is a 0 in the final column we consider the values in the row.

- If x value is 1 use x' in the sum term, if x value is 0 use x in the sum term.
- If y value is 1 use y' in the sum term, if y value is 0 use y in the sum term.
- If z value is 1 use z' in the sum term, if z value is 0 use z in the sum term.

We then multiply, or take the product of, all the sum terms we get. This is the product-of-sum form for the expression.

x	y	z	$x \times z' + y$	sum terms
1	1	1	1	
1	1	0	1	
1	0	1	0	$x' + y + z'$
1	0	0	1	
0	1	1	1	
0	1	0	1	
0	0	1	0	$x + y + z'$
0	0	0	0	$x + y + z$

In the third row the final column is 0 so we use this row. In this row the x value is 1, the y value is 0, and the z value is 1 so we use x', y, and z' in the sum term to get $x' + y + z'$. In the seventh row the final column is 0 so we use this row. In this row the x value is 0, the y value is 0, and the z value is 1 so we use x, y, and z' in the sum term to get $x + y + z'$. In the eighth row the final column is 0 so we use this row. In this row the x value is 0, the y value is 0, and the z value is 0 so we use x, y, and z in the sum term to get $x + y + z$. We then multiply, or take the product of, all the sum terms to get $x \times z' + y = (x' + y + z')(x + y + z')(x + y + z)$.

Example 5.14

Write the product-of-sums form for $[(x + y) \times z]'$.

x	y	z	$[(x + y) \times z]'$	sum term
1	1	1	0	$x' + y' + z'$
1	1	0	1	
1	0	1	0	$x' + y + z'$
1	0	0	1	
0	1	1	0	$x + y' + z'$
0	1	0	1	
0	0	1	1	
0	0	0	1	

So $[(x + y) \times z]' = (x' + y' + z')(x' + y + z')(x + y' + z')$.

5.5 PROBLEMS

Question 5.1 *(Easy)* Simplify the following Boolean algebra expressions.

(a) $x + x$
(b) $x + x'$
(c) $(x + x')'$
(d) $x + xy$
(e) $x + x'y$
(f) $x(x' + y)$
(g) $x(x)$
(h) $x(x + y)$
(i) $x(x + y + z)$
(j) $xy + xy'$
(k) $(x' + x')'$
(l) $xy + x'y$

Question 5.2 *(Moderate)* Simplify the following Boolean algebra expressions.

(a) $(x' + y')(x' + y)$
(b) $y + (yy')$
(c) $x' + yx'$
(d) $(x + y')(x + y)$
(e) $w + [w + (wx)]$
(f) $x[x + (xy)]$
(g) $w + (wx'yz)$
(h) $w'(wxyz)'$

Question 5.3 *(Difficult) Simplify the following Boolean algebra expressions.*

(a) $y[x + (x'y)]$

(b) $[(xy') + x']'$

(c) $x'y + x(x + y)$

(d) $(x + xy)'y$

(e) $xz' + x'y + (yz)'$

(f) $(xy)' + (yz)$

(g) $[x + (yz)](x' + z)$

(h) $xz' + x'y + (xz)'$

(i) $xz + x'y + zy$

(j) $x' + y' + xyz'$

(k) $(x + z)(x' + y)(z + y)$

(l) $(xy)'(x' + y)(y' + y)$

Question 5.4 *Write down the Boolean expression associated with each of the following circuits.*

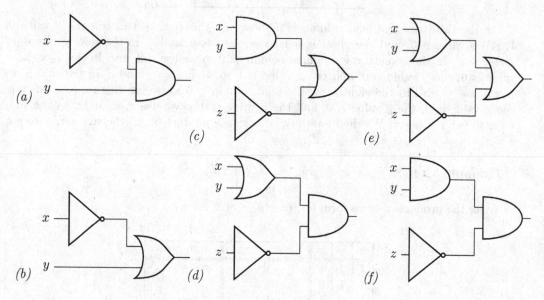

(a)

(b)

(c)

(d)

(e)

(f)

Question 5.5 *Write down the Boolean expression associated with each of the following circuits.*

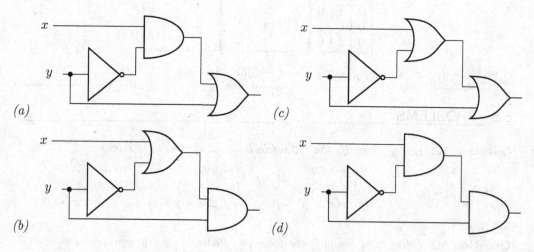

(a)

(b)

(c)

(d)

Question 5.6 *Write down the Boolean expression associated with each of the following circuits.*

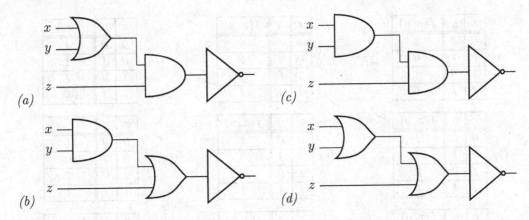

Question 5.7 *Write down the Boolean expression associated with each of the following circuits.*

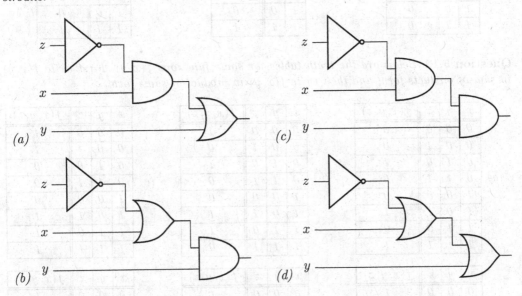

Question 5.8 *For the expressions in question 5.1, draw the associated circuit.*

Question 5.9 *For the expressions in question 5.2, draw the associated circuit.*

Question 5.10 *For the expressions in question 5.3, draw the associated circuit.*

Question 5.11 *Below are the truth tables for some functions $f(x, y)$. First write $f(x, y)$ in sum-of-products form and then write $f(x, y)$ in product-of-sums form.*

(a)

x	y	$f(x, y)$
0	0	1
0	1	0
1	0	1
1	1	0

(d)

x	y	$f(x, y)$
0	0	0
0	1	0
1	0	1
1	1	1

(g)

x	y	$f(x, y)$
0	0	1
0	1	1
1	0	1
1	1	0

(b)

x	y	$f(x, y)$
0	0	1
0	1	1
1	0	0
1	1	0

(e)

x	y	$f(x, y)$
0	0	1
0	1	0
1	0	0
1	1	1

(h)

x	y	$f(x, y)$
0	0	1
0	1	0
1	0	1
1	1	1

(c)

x	y	$f(x, y)$
0	0	0
0	1	1
1	0	0
1	1	0

(f)

x	y	$f(x, y)$
0	0	0
0	1	0
1	0	0
1	1	1

(i)

x	y	$f(x, y)$
0	0	0
0	1	1
1	0	0
1	1	1

Question 5.12 *Below are the truth tables for some functions $f(x, y)$. First write $f(x, y)$ in sum-of-products form and then write $f(x, y)$ in product-of-sums form.*

(a)

x	y	z	$f(x, y, z)$
0	0	0	0
0	0	1	0
0	1	0	1
0	1	1	0
1	0	0	0
1	0	1	1
1	1	0	0
1	1	1	1

(c)

x	y	z	$f(x, y, z)$
0	0	0	1
0	0	1	0
0	1	0	1
0	1	1	0
1	0	0	0
1	0	1	1
1	1	0	1
1	1	1	0

(e)

x	y	z	$f(x, y, z)$
0	0	0	1
0	0	1	1
0	1	0	0
0	1	1	0
1	0	0	0
1	0	1	1
1	1	0	1
1	1	1	1

(b)

x	y	z	$f(x, y, z)$
0	0	0	1
0	0	1	0
0	1	0	1
0	1	1	1
1	0	0	1
1	0	1	0
1	1	0	1
1	1	1	0

(d)

x	y	z	$f(x, y, z)$
0	0	0	1
0	0	1	0
0	1	0	0
0	1	1	1
1	0	0	0
1	0	1	0
1	1	0	1
1	1	1	0

(f)

x	y	z	$f(x, y, z)$
0	0	0	1
0	0	1	1
0	1	0	0
0	1	1	1
1	0	0	0
1	0	1	0
1	1	0	0
1	1	1	1

Question 5.13 *Write out the truth table for the following Boolean expressions.*

(a) $(x' + y)x$

(b) $(x + y')x$

(c) $(x + y)y'$

(d) $(x + y)(x + y)$

(e) $(x + y)(x' + y)$

(f) $(x + y')(x' + y)$

(g) $xx + x(x + y)$

(h) $xy + y(x + y)$

(i) $x'y + y'(x + y)$

(j) $x'y' + y'(x + y)$

(k) $xy(x + y)$

(l) $x'y'(x' + y')$

Question 5.14 *Write out the truth table for the following Boolean expressions.*

(a) $x(x + y' + z)$

(b) $x'(x + y' + z)$

(c) $(x + y)(x + z)$

(d) $(x' + y')(x + z)$

(e) $x'y' + z'(x + y)$

(f) $xy(x + y + z)$

CHAPTER 6

Functions

Functions play an important role in computer science, particularly in complexity analysis, which is used to analyze algorithms. An understanding of functions is also necessary for many topics in advanced computer science, meaning that computer science majors need to have a working understanding of functions.[1]

6.1 INTRODUCTION TO FUNCTIONS

You have probably already studied real-valued functions of a real-variable. When you studied these you probably thought of them as something with a formula, like $x^2 - 2x + 3$, or something you could graph using an x and y axes. But the real definition of functions is more general.

Definition 6.1 *Definition one of function: Let X and Y be sets. A **function** from X to Y is a rule that assigns to each element of X exactly one element of Y. A function f from X to Y is written as*

$$f : X \longrightarrow Y.$$

Lower case Latin letters or Greek letters are usually used to represent functions. If $f : X \to Y$ is a function from the set X to the set Y then the set X is called the **domain** of f and the set Y is called the **codomain** of f. If the rule f sends x in the domain to y in the codomain then y is called the **image** of x.

Example 6.1

Let the domain be $X = \{1, 2, 3\}$ and the codomain be $Y = \{a, b, c\}$. Then a function $f : X \to Y$ is defined by

$$f(1) = a,$$
$$f(2) = c,$$
$$f(3) = b.$$

The function f is a rule that assigns to $1 \in X$ the value $a \in Y$. This is what $f(1) = a$ means. The rule f assigns to $2 \in X$ the value $c \in Y$. That is what $f(2) = c$ means. And the rule f assigns to $3 \in X$ the value $b \in Y$. That is what $f(3) = b$ means. Another

---

[1]Note, the word "function" is often used in programming in a different way. There it refers to a block of reusable code in a program that can be called when it is needed.

way to say this is the image of 1 is a, the image of 2 is c, and the image of 3 is b. The function f is shown in a picture below.

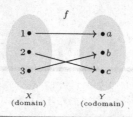

Example 6.2

Let the domain be $X = \{1, 2, 3\}$ and the codomain be $Y = \{a, b, c\}$. Then a function $f : X \to Y$ is defined by

$$f(1) = c,$$
$$f(2) = a,$$
$$f(3) = c.$$

Notice that this function has $f(1) = c$ and $f(3) = c$. According to the definition, a function is a rule that assigns to each element of X exactly one element of Y. For every element in X there is assigned exactly one element of Y. It is perfectly okay that both 1 and 3 are assigned to the same element c.

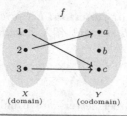

Example 6.3

Let the domain be $\mathbb{R}$ and the codomain be $\mathbb{R}$. Then a function $f : \mathbb{R} \to \mathbb{R}$ is defined by the rule

$$f(x) = 4x^2 - 10x.$$

Sometimes functions that are defined by an algebraic rule are shown using a picture of a function-machine.

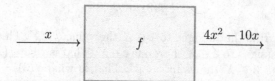

Example 6.4

Boolean expressions, and therefore circuits, can be viewed as multivariable functions. Consider the circuit

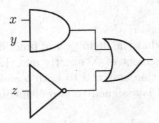

represented by the expression $(x \times y) + z'$. First recall that $\mathcal{B} = \{0, 1\}$. This circuit represents a function whose domain is the cartesian product $\mathcal{B} \times \mathcal{B} \times \mathcal{B}$ and whose codomain is $\mathcal{B}$,

$$f : \mathcal{B} \times \mathcal{B} \times \mathcal{B} \to \mathcal{B}.$$

The input of the function is an 3-tuple of Boolean variables $(x, y, z) \in \mathcal{B} \times \mathcal{B} \times \mathcal{B}$ and the output is an element of $\mathcal{B}$. The function rule is given by

$$f(x, y, z) = (x \times y) + z'.$$

We will not deal any further with multivariable functions in this class, but they are very common in both mathematics and computer science.

Example 6.5

Let the domain be $\mathbb{N}$ and the codomain be $Y = \{0, 1, 2, 3, 4, 5, 6, 7\}$. Then $f : \mathbb{N} \to Y$ where

$$f(x) = x \bmod 8$$

is a function. We have seen this function before. This function takes a natural number x and finds the remainder after division by 8.

Example 6.6

Let the domain be $X = \{$all circles in the plane$\}$ and the codomain be $\mathbb{R}$. Then $f : X \to \mathbb{R}$ where

$$f(x) = \text{the radius of } x$$

is a function. This function take a circle and finds the radius of the circle.

We can see that the idea of a function can apply to many different situations. This is what makes it such an important and useful concept. But before going further we want to consider an example of something that is not a function. Consider the following set relation:

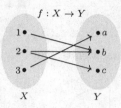

Is this a function? Remember what the definition of function says, a function from X to Y is a rule that assigns to each element of X exactly one element of Y. What is happening to the element $2 \in X$? It is getting assigned to both b and c. Thus $2 \in X$ is not assigned to exactly one element of Y, it is assigned to two elements of Y. Therefore, this is not a function.

Now let us reconsider the second example, the function $f : X \to Y$ given by

$$f(1) = c,$$
$$f(2) = a,$$
$$f(3) = c,$$

which we showed as

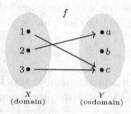

Notice that no element in X is send to $b \in Y$. The **image** of a function f is given by

$$\text{Im}(f) = \{y \in Y \mid y = f(x) \text{ for some } x \in X\}.$$

The image of a function is also called the **range** of the function. We will usually use the word range in this book. In this example the range of f is given by the set $\{a, c\}$. Here the range of f is not equal to the codomain of f. Instead, the range of f is a proper subset of the codomain of f,

$$\text{range of } f = \{a, c\} \subset \{a, b, c\} = \text{codomain of } f.$$

We always have range $\subseteq$ codomain. So, for each function there are three sets of interest, the domain, the codomain, and the range.

There is a second definition of function that uses the language of set theory. This is an important connection between set theory and functions. Recall, we studied set relations in chapter 4. Functions are a special kind of set relation. This is the way that computer scientists usually think about functions.

Definition 6.2 *Definition two of functions: Let X and Y be sets. A **function** $f : X \to Y$ is a subset of the Cartesian product of X and Y,*

$$f \subset X \times Y$$

such that

1. *If $x \in X$ then there exists a $y \in Y$ such that $(x, y) \in f$,*

2. *And this $y \in Y$ is unique. (That means if (x, y) and (x, z) are in f then $y = z$.)*

Example 6.7

Consider the function f shown below.

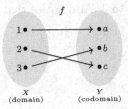

The domain is given by $X = \{1, 2, 3\}$ and the codomain is given by $Y = \{a, b, c\}$. Thus

$$X \times Y = \{(x, y) \mid x \in X, y \in Y\}$$
$$= \{(1, a), (2, a), (3, a), (1, b), (2, b), (3, b), (1, c), (2, c), (3, c)\}.$$

The function f is a subset of $X \times Y$ given by

$$f = \{(1, a), (2, c), (3, b)\} \subset X \times Y.$$

The first property in the definition is satisfied: if $x \in X$ then there is a $y \in Y$ such that $(x, y) \in f$. For $1 \in X$, we have $a \in Y$ such that $(1, a) \in f$. For $2 \in X$, we have $c \in Y$ such that $(2, c) \in f$. For $3 \in X$, we have $b \in Y$ such that $(3, b) \in f$. The second property is also satisfied: this $y \in Y$ is unique. For $1 \in X$ this $y \in Y$ is unique, namely $a \in Y$. For $2 \in X$ this $y \in Y$ is unique, namely $c \in Y$. For $3 \in X$ this $y \in Y$ is unique, namely $b \in Y$. Also notice that the range and the codomain are the same.

Example 6.8

Consider the function f shown below.

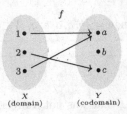

Like before the domain is given by $X = \{1, 2, 3\}$, the codomain is given by $Y = \{a, b, c\}$, and

$$X \times Y = \{(x, y) \mid x \in X, y \in Y\}$$
$$= \{(1, a), (2, a), (3, a), (1, b), (2, b), (3, b), (1, c), (2, c), (3, c)\}.$$

The function f is again subset of $X \times Y$,

$$f = \{(1, a), (2, c), (3, a)\} \subset X \times Y.$$

Convince yourself that the two properties in the definition are satisfied. Consider the second property. For $1 \in X$ the $y \in Y$ is unique, it is $a \in Y$ since $(1, a) \in f$. For $3 \in X$ the $y \in Y$ is unique, it is also $a \in Y$ since $(3, a) \in f$. It is okay that $a \in Y$ is used twice. The element $a \in Y$ is the unique element in Y associated to 1 and also the unique element in Y associated to 3. Also notice that

$$\text{range} = \{a, c\} \subset Y.$$

Example 6.9

Consider the function f shown below.

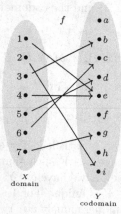

Here the domain is $X = \{1, 2, 3, 4, 5, 6, 7\}$ and the codomain is given by $Y = \{a, b, c, d, e, f, g, h, i\}$. We will not write out all 63 elements of $X \times Y$. The function f is given by

$$f = \{(1, e), (2, i), (3, b), (4, e), (5, d), (6, c), (7, g)\} \subset X \times Y.$$

Convince yourself that the two properties in the definition are satisfied. Also, notice that

$$\text{range} = \{b, c, d, e, g, i\} \subset Y.$$

Example 6.10

Recall our example of something that was not a function.

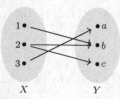

Here $X = \{1, 2, 3\}$, $Y = \{a, b, c\}$, and

$$X \times Y = \{(x, y) \mid x \in X, y \in Y\}$$
$$= \{(1, a), (2, a), (3, a), (1, b), (2, b), (3, b), (1, c), (2, c), (3, c)\}.$$

The set relation represented in this picture is given by

$$\{(1,b),(2,b),(2,c),(3,a)\} \subset X \times Y.$$

For $2 \in X$ we have $b \in Y$ such that $(2,b) \in f$. But we also have $c \in Y$ such that $(2,c) \in f$. Thus the second property is not satisfied. For $2 \in X$ the $y \in Y$ is not unique, there are two of them, namely b and c. Thus the definition for a function is not satisfied.

A function is called **onto** if the range of the function is equal to the codomain of the function. A function is called **one-to-one** if no two distinct elements of the domain have the same image. Another way of saying this is that for the elements x_1 and x_2 in the domain of f, if $x_1 \neq x_2$ then $f(x_1) \neq f(x_2)$. Now it is a good time to remember a little logic. If we define the propositions p and q to be

$$p = \text{``}x_1 \neq x_2\text{''} \qquad \text{and} \qquad q = \text{``}f(x_1) \neq f(x_2)\text{''}$$

then the statement "if $x_1 \neq x_2$ then $f(x_1) \neq f(x_2)$" is the expression $p \rightarrow q$. We showed that the expression $p \rightarrow q$ was logically equivalent to its contrapositive $\neg q \rightarrow \neg p$,

$$p \rightarrow q \equiv \neg q \rightarrow \neg p.$$

Therefore the statement "if $x_1 \neq x_2$ then $f(x_1) \neq f(x_2)$" is logically equivalent to the statements "if $f(x_1) = f(x_2)$ then $x_1 = x_2$." This is another way to define one-to-one functions.

Example 6.11

Examples of functions that are one-to-one and/or onto.

- Function is one-to-one and is onto.

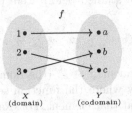

X (domain) Y (codomain)

- Function is one-to-one but is not onto.

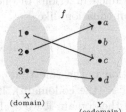

X (domain) Y (codomain)

- Function is not one-to-one but is onto.

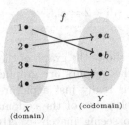

X (domain) Y (codomain)

- Function is not one-to-one and is not onto.

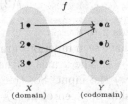

X (domain) Y (codomain)

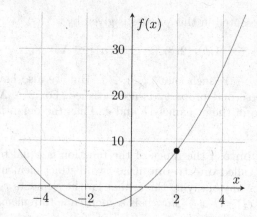

Figure 6.1 The graph of the function $f(x) = x^2 + 3x - 2$. This graph (gray) is exactly the set f defined by $\{(x, x^2 + 3x - 2) \mid x \in \mathbb{R}\} \subset \mathbb{R} \times \mathbb{R}$. The element $(2, 8) \in f$ is shown as a black dot. Every point on the graph (gray) represents an element of $f \subset \mathbb{R} \times \mathbb{R}$.

6.2 REAL-VALUED FUNCTIONS

In this section we will discuss real-valued functions of one variable. You have probably studied real-valued functions of one variable before. **Real-valued** means the codomain of the function is the set of real numbers, $\mathbb{R}$. **One variable** means the function only requires one variable as input. Here we will assume that this one variable comes from the real numbers $\mathbb{R}$ as well. Thus, a real-valued function of one variable is a fancy way of saying the domain and codomain of the function is $\mathbb{R}$, that is,

$$f : \mathbb{R} \longrightarrow \mathbb{R}.$$

The real-valued functions of one variable that you studied before were probably given in terms of an equation, like

$$f(x) = x^2 + 3x - 2 \qquad \text{or like} \qquad y = x^2 + 3x - 2.$$

The equation is the rule that assigns to each element $x \in \mathbb{R}$ (the domain) exactly one element of $\mathbb{R}$ (the codomain). If we want to know where the function sends a particular element of the domain $\mathbb{R}$ we evaluate the function at that element. For example, if we want to find where the function sends 2 we **evaluate** it at 2. That means we replace the x in the equation with 2,

$$f(2) = 2^2 + 3(2) - 2 = 4 + 6 - 2 = 8.$$

So f sends 2 to 8.

You have seen functions graphed before. Fig. 6.1 shows the graph of the function $f(x) = x^2 + 3x - 2$. The x-axis is the set of inputs and the outputs are graphed directly above (or below) the input. So, the graph of a function is really just a picture that shows the relationship between elements of the domain and elements of the codomain. It turns out that the graphs of functions that you are used to seeing match exactly the set theory definition of a function. Consider our function $f : \mathbb{R} \to \mathbb{R}$ defined by $f(x) = x^2 + 3x - 2$. According to the set theory definition of a function, a function is a subset $f \subset \mathbb{R} \times \mathbb{R}$ that satisfies two properties: (1) if $x \in \mathbb{R}$ then there is a $y \in \mathbb{R}$ such that $(x, y) \in f$, and (2) this

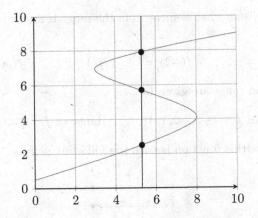

Figure 6.2 Using the vertical line test. The vertical line (black) intersects the graph (gray) in three places (black dots). Therefore the graph is not a function.

y is unique. Finally, recall that $\mathbb{R} \times \mathbb{R}$ is exactly the plane. For every input x from $\mathbb{R}$ there is an output y from $\mathbb{R}$ so that $(x, y) \in f \subset \mathbb{R} \times \mathbb{R}$. The output y is given by $x^2 + 3x - 2$, which is why we often write $y = x^2 + 3x - 2$ for a function. So, according to the definition the function f is given by

$$f = \{(x, x^2 + 3x - 2) \mid x \in \mathbb{R}\} \subset \mathbb{R} \times \mathbb{R}.$$

Every point in the graph of the function in Fig. 6.1 is an element of the function set $f \subset \mathbb{R} \times \mathbb{R}$. This definition gives us a wonderful way to tell if a graph is a function or not called the **vertical line test**. If the y is not unique at some value of x then there are at least two points (x, y_1) and (x, y_2), where $y_1 \neq y_2$, that are on the graph. Both of these points happen where the graph intersects a vertical line drawn at x. So, if it is possible to draw a vertical line that intersects the graph in two or more places then the graph does not represent a function, see Fig. 6.2.

Example 6.12

Find the domain and range of the graphed function and then find the function rule. Here the graph of the function is simply the four shown points.

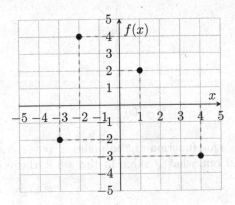

The function rule tells us what element from the range is assigned to an element from the domain,

$$f(-3) = -2, \qquad f(-2) = 4, \qquad f(1) = 2, \qquad f(4) = -3.$$

Thus the domain and range are given by

$$\text{Domain of } f = \{-3, -2, 1, 4\}, \qquad \text{Range of } f = \{-3, -2, 2, 4\}.$$

Another way to show this function is with the following picture.

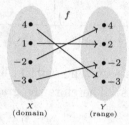

Finally, we could also have represented the function f as a subset of the domain $\times$ range,

$$f = \big\{(-3, -2), (-2, 4), (1, 2), (4, -3)\big\} \subset \text{Domain} \times \text{Range}.$$

Pay attention to the different ways we could write down a function. Make sure you understand how they are related.

Example 6.13

Consider the function $f : X \to \mathbb{R}$ given by

$$f(x) = \frac{5x}{2x - 1}.$$

The graph of this function is shown.

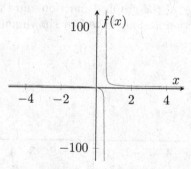

We start by evaluating this function at the points $x = -2$, $x = 0$, and $x = 6$. This means we want to find out what y value in the set $\mathbb{R}$ the values -2, 0, and 6 are assigned.

$$f(-2) = \frac{5(-2)}{2(-2)-1} = 2, \qquad f(0) = \frac{5(0)}{2(0)-1} = 0, \qquad f(6) = \frac{5(6)}{2(6)-1} = 2.727272\ldots$$

Thus -2 is assigned to, or sent to, 2, 0 is sent to 0, and 6 is sent to $2.727272\ldots$. Next we find the domain of f. For this we look carefully at the rule, or algebra expression, that defines f,

$$\frac{5x}{2x-1}.$$

We know that division by zero is undefined, so we are not allowed to use any x value that results in $2x - 1 = 0$. To find this x value we simply solve this equation to find $x = \frac{1}{2}$. Thus the domain is the set

$$\text{Domain of } f = \mathbb{R} - \left\{\frac{1}{2}\right\} = \left(-\infty, \frac{1}{2}\right) \cup \left(\frac{1}{2}, \infty\right).$$

Example 6.14

Use the vertical line test to determine if the following graph represents a function.

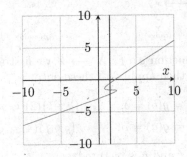

Since it is possible to draw a vertical line that intersects the graph (gray) more than once, the graph does not represent a function.

Example 6.15

Use the vertical line test to determine if the following graph represents a function.

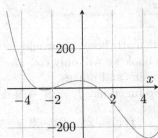

Since it is not possible to draw a vertical line that intersects the graph (gray) more than once, the graph does represent a function.

6.3 FUNCTION COMPOSITION AND INVERSES

We will now study a way to link two or more functions together called composition. When we compose two functions the output of the first function becomes the input of the second function.

Definition 6.3 *Suppose we have two functions, $f : X \longrightarrow Y$ and $g : Y \longrightarrow Z$. The* ***composition*** *of f and g is a new function $g \circ f : X \longrightarrow Z$ given by $(g \circ f)(x) = g\big(f(x)\big)$.*

The new function $g \circ f$ is sometimes called the **composite function** of f and g. Notice that when we write $g \circ f$ we apply f first then then apply g to the result.

Example 6.16

Suppose we are given the functions $f : X \to Y$ and $g : Y \to Z$ as pictured. Find $g \circ f$.

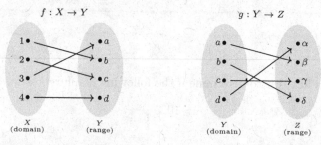

To find the composition function $g \circ f : X \longrightarrow Z$ we first apply f to the elements of X, find their outputs, and then apply g to these outputs,

$$(g \circ f)(1) = g\big(f(1)\big) = g(b) = \delta, \qquad (g \circ f)(2) = g\big(f(2)\big) = g(c) = \gamma,$$
$$(g \circ f)(3) = g\big(f(3)\big) = g(a) = \beta, \qquad (g \circ f)(4) = g\big(f(4)\big) = g(d) = \alpha.$$

The composite function of f and g is pictured as

$f : X \to Y \qquad g : Y \to Z$

1	α
2	β
3	γ
4	δ

X (domain)   Y (range/domain)   Z (range)

If this were a more advanced math book we would have to be very careful with our domains and ranges. But in this book we will only consider simple functions where we will not need to worry about the domains and ranges.

Now suppose we are given two functions, $f, g : \mathbb{R} \to \mathbb{R}$ defined by

$$f(x) = x^2 + 2x - 3$$

and

$$g(x) = 3x - 1.$$

We want to find the composite function $g \circ f$. One can imagine this as a series of function machines. Here the variable x enters the function machine f and the output is $f(x)$, which then enters the next function machine g. The output from function machine g is $g(f(x))$.

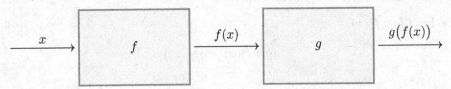

If we put $x = 4$ into the function machine f then $f(4) = 4^2 + 2(4) - 3 = 21$ would come out. Then the 21 would go into the g machine and $g(21) = 3(21) - 1 = 62$ would come out. The same thing would happen for any other number you put in.

Now we find the formula for the composite function by putting the variable x into the machine instead of a number. Clearly $x^2 + 2x - 3$ would come out of the f machine which would then go into the g machine. What would come out of the g machine? To find out we put the $x^2 + 2x - 3$ into the formula for $g(x)$ and then simplify,

$$
\begin{aligned}
(g \circ f)(x) &= g(f(x)) \\
&= g(x^2 + 2x - 3) \\
&= 3(x^2 + 2x - 3) - 1 \\
&= 3x^2 + 6x - 9 - 1 \\
&= 3x^2 + 6x - 10.
\end{aligned}
$$

This gives us a formula for $(g \circ f)(x) = 3x^2 + 6x - 10$. Notice that

$$
\begin{aligned}
(g \circ f)(4) &= 3(4)^2 + 6(4) - 10 \\
&= 3(16) + 24 - 10 \\
&= 62,
\end{aligned}
$$

which is exactly what we got before.

To find $f \circ g$ we do the same thing. We can imagine a series of function machines.

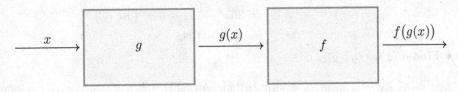

Then to find $f \circ g$ we put the formula for $g(x)$ into the formula for $f(x)$ and simplify,

$$
\begin{aligned}
(f \circ g)(x) &= f(g(x)) \\
&= f(3x - 1) \\
&= (3x - 1)^2 + 2(3x - 1) - 3 \\
&= (9x^2 - 6x + 1) + (6x - 2) - 3 \\
&= 9x^2 - 4.
\end{aligned}
$$

This gives the formula for $(f \circ g)(x) = 9x^2 - 4$. Notice, $g \circ f \neq f \circ g$. Of course we can compose more than two functions. We can compose three functions, or even more.

Example 6.17

Let $f(x) = 2x - 3$, $g(x) = -3x$, and $h(x) = x^2 + 3$.

- Find $(g \circ f)(x)$.

$$\begin{aligned}
(g \circ f)(x) &= g\big(f(x)\big) \\
&= g\big(2x - 3\big) \\
&= -3(2x - 3) \\
&= -6x + 9
\end{aligned}$$

- Find $(f \circ g)(x)$.

$$\begin{aligned}
(f \circ g)(x) &= f\big(g(x)\big) \\
&= f\big(-3x\big) \\
&= 2(-3x) - 3 \\
&= -6x - 3
\end{aligned}$$

- Find $(h \circ f)(x)$.

$$\begin{aligned}
(h \circ f)(x) &= h\big(f(x)\big) \\
&= h\big(2x - 3\big) \\
&= (2x - 3)^2 + 3 \\
&= 4x^2 - 12x + 9 + 3 \\
&= 4x^2 - 12x + 12
\end{aligned}$$

- Find $(h \circ g \circ f)(x)$. We already know $g\big(f(x)\big) = -6x + 9$

$$\begin{aligned}
(h \circ g \circ f)(x) &= h\Big(g\big(f(x)\big)\Big) \\
&= h\big(-6x + 9\big) \\
&= (-6x + 9)^2 + 3 \\
&= 36x^2 - 54x - 54x + 81 + 3 \\
&= 36x^2 - 108x + 84
\end{aligned}$$

- Find $g^2(x) = (g \circ g)(x)$.

$$\begin{aligned}
(g \circ g)(x) &= g\big(g(x)\big) \\
&= g\big(-3x\big) \\
&= -3(-3x) \\
&= 9x.
\end{aligned}$$

Positive exponents on functions are different from exponents on numbers. Instead of meaning repeated multiplications they mean repeated compositions.

We now will study how to find the inverse of a function. Again, in a more advanced book we would have to be very careful with domains and ranges, but here all of our examples will be very simple. We begin by defining the identity function and the inverse function.

Definition 6.4 *Suppose X is a set. The **identity function** on X is the function $i : X \to X$ given by $i(x) = x$.*

Definition 6.5 *Suppose $f : X \to Y$ and $g : Y \to X$ are functions. If $g \circ f : X \to X$ is the identity function on X and $f \circ g : Y \to Y$ is the identity function on Y then g is the **inverse function** of f and f is the inverse function of g.*

The inverse function of f is usually written as f^{-1}. The exponent -1 tells you we are talking about the inverse function of f. It works differently than negative exponents on numbers. Also, if f is not a one-to-one function then f does not have an inverse.

Example 6.18

If $X = \{1, 2, 3, 4\}$ then the identity function on X is given by the rule

$$i(1) = 1, \qquad i(2) = 2, \qquad i(3) = 3, \qquad i(4) = 4.$$

The identity function on X is shown here.

$i : X \to X$

X (domain) X (codomain)

Example 6.19

If $X = \{1, 2, 3\}$, $Y = \{a, b, c\}$, find the inverse function of f for $f : X \to Y$ given by

$$f(1) = a, \qquad f(2) = c, \qquad f(3) = b.$$

Function f is pictured by

$f : X \to Y$

X (domain) Y (codomain)

The inverse function $f^{-1} : Y \to X$ is given by

$$f^{-1}(a) = 1, \qquad f^{-1}(c) = 2, \qquad f^{-1}(b) = 3.$$

You can get the picture for f^{-1} just by reversing the arrows on the picture for f. This give us

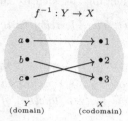

$$f^{-1} : Y \to X$$

Y (domain) $\qquad$ X (codomain)

Notice that

$$(f^{-1} \circ f)(1) = f^{-1}(f(1)) = f^{-1}(a) = 1,$$
$$(f^{-1} \circ f)(2) = f^{-1}(f(2)) = f^{-1}(c) = 2,$$
$$(f^{-1} \circ f)(3) = f^{-1}(f(3)) = f^{-1}(b) = 3,$$

so $f^{-1} \circ f$ is exactly the identity function i.

Example 6.20

Let $f, g : \mathbb{R} \to \mathbb{R}$ be given by $f(x) = x + 6$ and $g(x) = x - 6$. Show that g is the inverse function of f.

$$\xrightarrow{\quad x \quad} \boxed{\quad f \quad} \xrightarrow{\quad f(x) \quad} \boxed{\quad g \quad} \xrightarrow{\quad g(f(x)) = i(x) \quad}$$

But that is easy to do since

$$(g \circ f)(x) = g(f(x))$$
$$= g(x + 6)$$
$$= (x + 6) - 6$$
$$= x$$

and so $g \circ f = i$, the identity function. In other words, we have found that $f^{-1}(x) = x - 6$.

Example 6.21

Let $f : \mathbb{R} \to \mathbb{R}$ be given by

$$f(x) = \frac{3x - 2}{5}.$$

To find $f^{-1} : \mathbb{R} \to \mathbb{R}$ we first rewrite the function and call the outputs y,

$$y = \frac{3x - 2}{5}.$$

We then switch the variables x and y in the equation,

$$x = \frac{3y - 2}{5}.$$

We then solve for y,

$$x = \frac{3y - 2}{5}$$
$$\Rightarrow \quad 5x = 3y - 2$$
$$\Rightarrow \quad 3y = 5x + 2$$
$$\Rightarrow \quad y = \frac{5x + 2}{3}.$$

We have found the inverse function, $f^{-1}(x) = \frac{5x+2}{3}$.

6.4 PROBLEMS

Question 6.1 *For the function pictured in Fig. 6.3,*

(a) *find the domain, codomain, and range of the function,*
(b) *determine if the function is one-to-one and/or onto,*
(c) *write the function as a rule that acts on each element of the domain, and*
(d) *write the function as a subset of the cartesian product of the domain and the codomain.*

Question 6.2 *Let the domain be $X = \{1, 2, 3, 4\}$ and the codomain be $Y = \{a, b, c, d, e\}$. The functions f_1, f_2, and f_2 are given by the rules:*

$f_1(1) = c$	$f_2(1) = c$	$f_3(1) = b$
$f_1(2) = b$	$f_2(2) = d$	$f_3(2) = b$
$f_1(3) = c$	$f_2(3) = e$	$f_3(3) = d$
$f_1(4) = e$	$f_2(4) = b$	$f_3(4) = d$

For each of these functions,

(a) *find the range of the function,*
(b) *determine if the function is one-to-one and/or onto,*
(c) *draw a picture showing how the function that acts on each element of the domain, and*
(d) *write the function as a subset of the cartesian product of the domain and the codomain.*

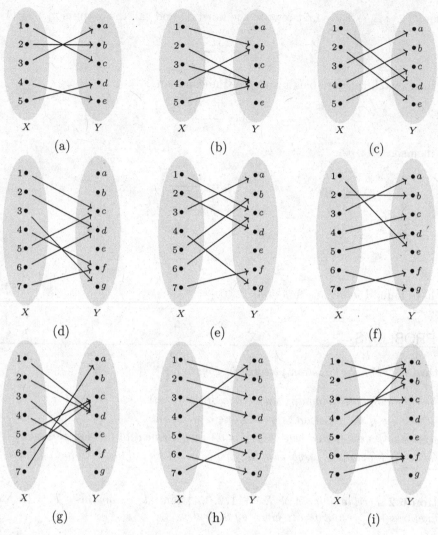

Figure 6.3 Functions for questions 6.1 and 6.11.

Question 6.3 *Let the domain be* $X = \{1, 2, 3, 4, 5, 6, 7\}$ *and the codomain be* $Y = \{a, b, c, d, e, f\}$. *The functions* f_1, f_2, *and* f_3 *are given by the rules:*

$f_1(1) = c$	$f_2(1) = a$	$f_3(1) = b$
$f_1(2) = d$	$f_2(2) = c$	$f_3(2) = c$
$f_1(3) = e$	$f_2(3) = e$	$f_3(3) = d$
$f_1(4) = f$	$f_2(4) = b$	$f_3(4) = e$
$f_1(5) = a$	$f_2(5) = d$	$f_3(5) = d$
$f_1(6) = b$	$f_2(6) = f$	$f_3(6) = c$
$f_1(7) = f$	$f_2(7) = d$	$f_3(7) = b$

For each of these functions,

(a) find the range of the function,

(b) determine if the function is one-to-one and/or onto,

(c) draw a picture showing how the function that acts on each element of the domain, and

(d) write the function as a subset of the cartesian product of the domain and the codomain.

Question 6.4 *Let the domain be* $X = \{1, 2, 3, 4, 5\}$ *and the codomain be* $Y = \{a, b, c, d, e\}$. *The functions* f_1, f_2, *and* f_3 *are given as subsets of the Cartesian product of* X *and* Y *by:*

$$f_1 = \Big\{(1, c), (2, b), (3, d), (4, c), (5, a)\Big\} \subset X \times Y$$

$$f_2 = \Big\{(1, d), (2, d), (3, c), (4, b), (5, a)\Big\} \subset X \times Y$$

$$f_3 = \Big\{(1, a), (2, d), (3, c), (4, b), (5, e)\Big\} \subset X \times Y$$

For each of these functions,

(a) *find the range of the function,*
(b) *determine if the function is one-to-one and/or onto,*
(c) *draw a picture showing how the function acts on each element of the domain, and*
(d) *write the function as rule.*

Question 6.5 *Let the domain be* $X = \{1, 2, 3, 4, 5, 6\}$ *and the codomain be* $Y = \{a, b, c, d, e, f\}$. *The functions* f_1, f_2, *and* f_3 *are given as subsets of the Cartesian product of* X *and* Y *by:*

$$f_1 = \Big\{(1, b), (2, c), (3, a), (4, e), (5, f), (6, d)\Big\} \subset X \times Y$$

$$f_2 = \Big\{(1, b), (2, b), (3, b), (4, e), (5, e), (6, e)\Big\} \subset X \times Y$$

$$f_3 = \Big\{(1, c), (2, d), (3, e), (4, f), (5, a), (6, b)\Big\} \subset X \times Y$$

For each of these functions,

(a) *find the range of the function,*
(b) *determine if the function is one-to-one and/or onto,*
(c) *draw a picture showing how the function acts on each element of the domain, and*
(d) *write the function as a rule.*

Question 6.6 *For the functions graphed in Fig. 6.4,*

(a) *find the domain and range of the function,*
(b) *draw a picture showing how the function acts on each element of the domain,*
(c) *write the function as a rule that acts on each element of the domain, and*
(d) *write the function as a subset of the cartesian product of the domain and the codomain.*

Question 6.7 *The following function rules are given by an equation. Evaluate these functions at the values* $x = -3$, $x = 0$, *and* $x = 4$.

(a) $f(x) = 3x + 7$
(b) $f(x) = 3(x + 4) - 5$
(c) $f(x) = -4x + 6$
(d) $f(x) = 2^x$
(e) $f(x) = x^2 + 3x$
(f) $f(x) = (x - 2)(x + 3)$
(g) $f(x) = -x^3 + 2x^2 - x$
(h) $f(x) = 2 \cdot 3^{-x+1}$
(i) $f(x) = (x^2 - 1)^2$

Question 6.8 *Use the vertical line test to determine if the graphs in Fig. 6.5 represent functions. If they do, determine the domain and range of the functions.*

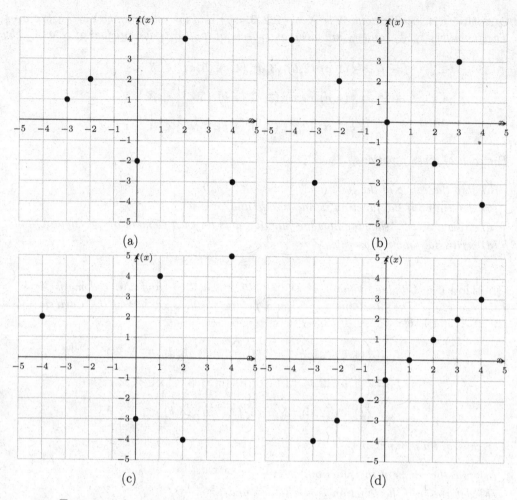

Figure 6.4 Functions for question 6.6.

Question 6.9 *Let* $X = \{1, 2, 3, 4\}$ *and* $Y = \{a, b, c, d\}$ *and* $Z = \{\alpha, \beta, \gamma, \delta\}$. *The functions* $f : X \rightarrow Y$ *and* $g : Y \rightarrow Z$ *are given by*

$$
\begin{aligned}
f(1) &= c & g(a) &= \delta \\
f(2) &= b & g(b) &= \gamma \\
f(3) &= d & g(c) &= \alpha \\
f(4) &= a & g(d) &= \beta
\end{aligned}
$$

Find $g \circ f : X \rightarrow Z$

Question 6.10 *Given the functions* f *and* g *in Fig. 6.6, find* $g \circ f$. *What is the domain, codomain, and range of* $g \circ f$? *Then find* $f \circ g$. *What is the domain, codomain, and range of* $f \circ g$?

Question 6.11 *For each function pictured in Fig. 6.3, find the inverse function if it exits.*

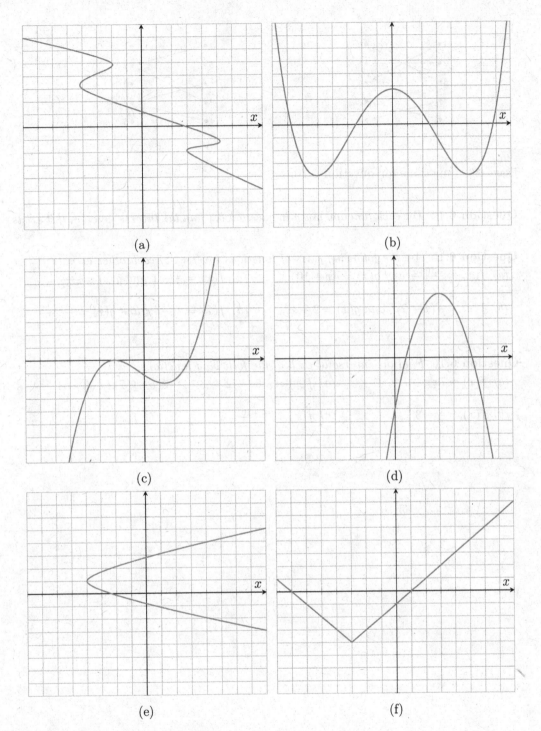

Figure 6.5 Graphs for question 6.8.

Question 6.12 *For each function given in question 6.2 find the inverse function if it exists.*

Question 6.13 *For each function given in question 6.3 find the inverse function if it exists.*

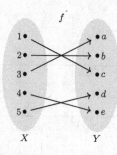

Figure 6.6 Functions for question 6.10.

Question 6.14 *For each function given in question 6.4 find the inverse function if it exists.*

Question 6.15 *Are the following pairs of functions inverses of each other or not?*

(a) $f(x) = \frac{-10+x}{4}$ and $g(x) = 4x + 10$

(c) $f(x) = \frac{3}{x} - 1$ and $g(x) = \frac{3}{x+1}$

(b) $f(x) = -(x+2)^3$ and $g(x) = 2 + x^3$

(d) $f(x) = 2 - \frac{2}{3}x$ and $g(x) = \frac{-2x+4}{3}$

Question 6.16 *Find the inverse of each of the following functions.*

(a) $f(x) = 7x$

(c) $f(x) = -5x + 2$

(e) $f(x) = \sqrt{x} - 4$

(b) $f(x) = \frac{1}{x} - 3$

(d) $f(x) = \frac{8x+10}{3}$

(f) $f(x) = \frac{2}{x-3}$

CHAPTER 7

Counting and Combinatorics

The basic concepts of counting and combinatorics help us understand a wide range of different situations. In computer science they help us understand such things as how many rows a truth table needs, how many different subsets a set can have, the maximum possible number of times a loop is executed in an algorithm, and how many possible spanning trees a graph has.

7.1 ADDITION AND MULTIPLICATION PRINCIPLES

Throughout this course we have already, and will continue to, encounter a few different counting problems. For example, given three propositions p, q, and r, each of which is either true or false, how many possible combinations of true and false are there for these three propositions? This will give us the number of rows we need in making a truth table. Or we want to know how many possible subsets a set A could have. This gives us the cardinality of the power set of A. Or suppose we want to know how many possible spanning trees a graph has. Or the maximum possible number of times a loop in an algorithm could be executed. These are all examples of counting problems in computer science. When it comes to counting there are two very important ideas, or principles, that get used. They are the addition principle and the multiplication principle. We will begin with the addition principle.

Definition 7.1 *The Addition Principle: Suppose there are n_1 ways for event E_1 to occur, and n_2 ways for event E_2 to occur, and we cannot do both E_1 and E_2. If all these ways are distinct then the number of ways for E_1 or E_2 to occur is $n_1 + n_2$.*

When we say that all the ways are distinct, we mean that none of the ways are the same as any of the other ways. Let us look at an example to see how this works.

Example 7.1

Suppose you go to a bakery and want either a chocolate cake or a vanilla cake. The bakery has 10 different chocolate cakes and 8 different vanilla cakes to choose from. How many choices of cake do you have?

Let E_1 be the event of buying a chocolate cake and E_2 be the event of buying a vanilla cake. Then clearly there are $n_1 = 10$ ways for E_1 to occur and $n_2 = 8$ ways for E_2 to occur. There are $n_1 + n_2 = 10 + 8 = 18$ ways for E_1 or E_2 to occur. Thus you have 18 choices for buying a chocolate cake or buying a vanilla cake.

Using the language of set theory, the addition principle could also be written this way.

Definition 7.2 *The Addition Principle in the language of set theory: If A and B are finite sets with $A \cap B = \varnothing$, then*

$$|A \cup B| = |A| + |B|.$$

Recall that $|A|$ mean the cardinality of set A, that is, the number of elements contained in set A. The addition principle can of course be generalized to more than two events. But before giving that definition we need to spend a moment to make sure we understand some notation. The capital Greek letter sigma, written as Σ, means "sum" or add. Suppose we had the formula

$$\sum_{i=1}^{4} x_i.$$

Notice the $i = 1$ written below the Σ and the 4 is written above the Σ. What exactly does this formula mean? It means we are to add all the variables x_i from $i = 1$ to 4. This is very similar to how the **for-do** loops work. So we have

$$\sum_{i=1}^{4} x_i = x_1 + x_2 + x_3 + x_4.$$

Similarly, the large $\bigcup$ means to take the union. Suppose we had the formula

$$\bigcup_{i=1}^{6} A_i.$$

This works exactly like above, it means we take the union of all the sets A_i from $i = 1$ to 6. Thus,

$$\bigcup_{i=1}^{6} A_i = A_1 \cup A_2 \cup A_3 \cup A_4 \cup A_5 \cup A_6.$$

Of course there is nothing special about 4 or 6. It could be any number at all. Now we are ready for the next definition.

Definition 7.3 *The generalized Addition Principle: If $A_1, A_2, \ldots, A_n$ are finite sets with $A_i \cap A_j = \varnothing$ for all i, j with $1 \leq i < j \leq n$, then*

$$|A_1 \cup A_2 \cup \cdots \cup A_n| = |A_1| + |A_2| + \cdots + |A_n|,$$

which could also be written as

$$\left| \bigcup_{i=1}^{n} A_i \right| = \sum_{i=1}^{n} |A_i|.$$

Example 7.2

Suppose you want to go from town X to town Y. You can go by air, by land, or by sea. There are two different ways to go by air, four different ways to go by land, and three ways to go by sea. How many ways are there to go from town X to town Y?

If A_1 is the set that contains the ways to go by air then $|A_1| = 2$. If A_2 is the set that contains the ways to go by land then $|A_2| = 4$. If A_3 is the set that contains the ways to go by sea then $|A_3| = 3$. Clearly $A_1 \cap A_2 = \varnothing$, $A_1 \cap A_3 = \varnothing$, and $A_2 \cap A_3 = \varnothing$. So we have

$$
\begin{aligned}
|A_1 \cup A_2 \cup A_3| &= |A_1| + |A_2| + |A_3| \\
&= 2 + 4 + 3 \\
&= 9.
\end{aligned}
$$

Now we will take a look at the multiplication principle.

Definition 7.4 *The Multiplication Principle: Suppose there are n_1 ways for event E_1 to occur, and each possible way that E_1 occurs allows for exactly n_2 ways for event E_2 to occur. Then the number of ways for E_1 **and** E_2 to occur is $n_1 \cdot n_2$.*

Example 7.3

In a class of 11 boys and 12 girls, how many ways are there to select two students, one a boy and one a girl?

Let E_1 be the event of choosing one boy, and E_2 be the event of choosing one girl. There are $n_1 = 11$ ways for E_1 to occur and $n_2 = 12$ ways for E_2 to occur. There are $n_1 \cdot n_2 = 11 \cdot 12 = 132$ ways for E_1 and E_2 to occur. Thus there are 132 ways to select one boy and one girl from the class.

Using the language of set theory, the multiplication principle could also be written this way.

Definition 7.5 *The Multiplication Principle in the language of set theory: If A and B are finite sets and $A \times B$ is the Cartesian product of A and B, then*

$$
|A \times B| = |A| \cdot |B|.
$$

Example 7.4

You own 7 pairs of pants and 15 shirts. How many outfits can you make?

Let A be the set that contains 7 pairs of pants and B be the set that contains 15 shirts. Each outfit consists of one pair of pants and one shirt, so each outfit is an element of $A \times B$. The number of different outfits possible is

$$
|A \times B| = |A| \cdot |B| = 7 \cdot 15 = 105.
$$

The multiplication principle can also be generalized to more than two events.

Definition 7.6 *The generalized Multiplication Principle: If $A_1, A_2, \ldots, A_n$ are finite sets then*

$$\left| A_1 \times A_2 \times \cdots \times A_n \right| = |A_1| \cdot |A_2| \cdots |A_n|.$$

Example 7.5

How many different seven digit binary sequences can be formed?

Here we have $A_i = \{0, 1\}$ for $1 \leq i \leq 7$. A seven digit binary sequence is an element of the set $A_1 \times A_2 \times \cdots \times A_7$. The number of different seven digit binary sequences is given by

$$\left| A_1 \times A_2 \times \cdots \times A_7 \right| = |A_1| \cdot |A_2| \cdots |A_7| = 2^7 = 128.$$

7.2   COUNTING ALGORITHM LOOPS

One of the most important applications of the addition principle and the multiplication principle in computer science is in counting the maximum number of times conditional controls and loops are executed in algorithms. This is very important in helping computer scientists and programmers know how long their algorithms will take to execute. We will cover this in chapter 8. For now we will simply use the addition and multiplication principles to see how they can help us.

Example 7.6

What is the maximum number of times the **while-do** loop in step 3 is executed?

Algorithm: Evaluates $n!$ where n is a positive integer.

1. Input positive integer n.
2. *answer* $\longleftarrow n$
3. **While** $n > 1$ **do**
    3.1 $n \longleftarrow n - 1$
    3.2 *answer* $\longleftarrow$ *answer* $\times n$
4. Output n.

Strictly speaking, this example does not require either the addition or the multiplication principles. The **while-do** loop in step 3 executes while $n > 1$. Notice that in step 3.1 we have $n \longleftarrow n - 1$, so n is decreased by one each time the loop executes. Suppose we input positive integer $n = 5$; then the **while-do** loop would execute when n was $5, 4, 3$, and 2. It would not execute when $n = 1$. Therefore the **while-do** loop would execute four times. Suppose we input positive integer $n = 10$. This time it would execute nine times. In general, whenever we input the positive integer n the **while-do** loop executes $n - 1$ times.

Example 7.7

What is the maximum number of times the **for-do** loop in step 2 is executed? What is the maximum number of times the **if-then** conditional control in step 2.1 is executed?

Algorithm: Searches a string of integers $x_1, \ldots, x_n$ to see if it contains the integer s.

1. Input string $x_1, \ldots, x_n$, n and s.
2. **For** $i = 1$ to n **do**
   2.1. **If** $x_i = s$ **then**
      2.1.1. Return "String contains s."
3. Return "String does not contain s."

Again, strictly speaking, this example does not require either the addition or the multiplication principles. However, it shows us something important. The **for-do** loop in step 2 is executed when i goes from 1 to n. This means it is executed a maximum of n times. Each time the **for-do** loop is executed the **if-then** conditional control is also executed. Thus the maximum number of times the **if-then** conditional control is executed is n times.

Notice, it is possible that both the **for-do** loop and **if-then** conditional control are executed fewer times. If for some number $i < n$ we have $x_i = s$ then the **if-then** conditional control step 2.1.1 is executed, the phrase "String contains s" is returned, and the algorithm ends after executing both the **for-do** loop and **if-then** conditional control only i times. But since we do not know what string $x_1, \ldots, x_n$ was input we cannot say for sure if the **for-do** loop will execute less than n times. Computer scientists and programmers are most interested in the maximum number of times a loop or conditional control is executed.

Example 7.8

What is the maximum number of times the **for-do** loops in steps 2 and 3 are executed? What is the maximum number of times the **if-then** conditional controls in steps 2.1 and 3.1 are executed?

Algorithm: Searches two strings of integers $x_1, x_2, \ldots, x_n$ and $y_1, y_2, \ldots, y_n$ of equal length to see if either string contains integer s.

1. Input strings $x_1, \ldots, x_n$ and $y_1, \ldots, y_n$, n and s.
2. **For** $i = 1$ to n **do**
   2.1. **If** $x_i = s$ **then**
      2.1.1. Return "String one contains integer s."
3. **For** $i = 1$ to n **do**
   3.1. **If** $y_i = s$ **then**
      3.1.1. Return "String two contains integer s."
4. Return "Neither string contains integer s."

The **for-do** loop in step 2 is executed when i goes from 1 to n. This means that the maximum number of times the **for-do** loop is executed is n times. Similarly, the **for-do** loop in step 3 is executed when i goes from 1 to n, so the maximum number of times this loop is executed is n. We use the addition principle to find the maximum number of times the **for-do** loops in both steps 2 and 3 are executed; n times in step 2 and n times in step 3 for a total of $n + n = 2n$ times.

It is also clear that the **if-then** conditional control in step 2.1 is executed once for each time the **for-do** loop in step 2 is executed. Similarly, the **if-then** conditional control in step 3.1 is executed once for each time the **for-do** loop in step 3 is executed. Using the addition principle, the maximum number of times the **if-then** conditional control in steps 2.1 and 3.1 are executed is $n + n = 2n$ times.

Example 7.9

What is the maximum number of times the **for-do** loop in step 2 is executed? What is the maximum number of times the **for-do** loop in step 2.1 is executed? What is the maximum number of times the **if-then** conditional control in step 2.1.1 is executed?

> Algorithm: Checks if two strings $x_1, \ldots, x_n$ and $y_1, \ldots, y_n$ contain a common integer. (That is, check and see if there is an integer s that is contained in both strings.)
>
> 1. Input strings $x_1, \ldots, x_n$ and $y_1, \ldots, y_n$.
> 2. For $i = 1$ to n **do**
>   2.1 For $j = 1$ to n **do**
>   2.1.1 If $x_i = y_j$ **then**
>   2.1.1.1 Return "There is an element common to both strings."
> 3. Return "There are no elements in common to both strings."

The maximum number of times the **for-do** loop in step 2 is executed is clearly n times. To figure out the maximum number of times the **for-do** loop in step 2.1 is executed requires the multiplication principle. When $i = 1$ the **for-do** loop in step 2.1 is executed n times. Then when $i = 2$ the **for-do** loop in step 2.1 is again executed n times. The same when $i = 3$, and so on. So for each of the n times the **for-do** loop in step 2 is executed the **for-do** loop in step 2.1 is executed n times. This means the **for-do** loop in step 2.1 is executed a maximum of $n \times n = n^2$ times. The **if-then** conditional control in step 2.1.1 is executed once for each time the **for-do** loop in step 2.1 is executed, so it is also executed a maximum of n^2 times.

7.3 PERMUTATIONS AND ARRANGEMENTS

Often we need to know how many different ways can we order, or arrange, a number of items. For example, how many different ways are there to order the three letters a, b, and c? A little thought will give us

$$abc, \quad acb, \quad bac, \quad bca, \quad cab, \quad cba.$$

Each ordering is called a **permutation** of the letters a, b, and c. There are a total of six **permutations** of the letters a, b, and c. One way to think about this is that there are three letters, so we need three spaces, one for each letter. In the first space we can put any of the three letters. In the second space we can put either of the two remaining letters, and in the third space we put whatever letter is left,

$$\underbrace{}_{\substack{3 \\ \text{possibilites}}} \quad \underbrace{}_{\substack{2 \\ \text{possibilites}}} \quad \underbrace{}_{\substack{1 \\ \text{possibility}}}$$

Finding out how many different permutations are possible becomes an application of the multiplication principle. There are a total of $3 \cdot 2 \cdot 1 = 6$ permutations possible.

We will introduce a little bit of notation to help us called the **factorial** symbol. The factorial symbol is simply the exclamation mark ! following a positive integer. By looking at a few examples it should be clear how the factorial sign works

$$1! = 1,$$
$$2! = 2 \cdot 1 = 2,$$
$$3! = 3 \cdot 2 \cdot 1 = 6,$$
$$4! = 4 \cdot 3 \cdot 2 \cdot 1 = 24,$$
$$5! = 5 \cdot 4 \cdot 3 \cdot 2 \cdot 1 = 120,$$
$$6! = 6 \cdot 5 \cdot 4 \cdot 3 \cdot 2 \cdot 1 = 720,$$
$$7! = 7 \cdot 6 \cdot 5 \cdot 4 \cdot 3 \cdot 2 \cdot 1 = 5\,040,$$
$$8! = 8 \cdot 7 \cdot 6 \cdot 5 \cdot 4 \cdot 3 \cdot 2 \cdot 1 = 40\,320,$$
$$9! = 9 \cdot 8 \cdot 7 \cdot 6 \cdot 5 \cdot 4 \cdot 3 \cdot 2 \cdot 1 = 362\,880.$$

Even if it looks a little funny we also define $0! = 1$. In general we have

$$n! = n \cdot (n-1) \cdot (n-2) \cdots 3 \cdot 2 \cdot 1.$$

We have actually encountered factorials before when we were looking at algorithms. Here is one possible algorithm for computing factorials.

Algorithm: Evaluates $n!$ where n is a positive integer.

1. Input positive integer n.
2. *answer* ⟵ n
3. **While** $n > 1$ **do**
   3.1 n ⟵ $n - 1$
   3.2 *answer* ⟵ *answer* $\times n$
4. Output n.

Factorials are exactly the notation we need to help us write down how many permutations are possible. The number of permutations of a set of n elements is given by $n!$.

Example 7.10

How many permutations of the letters p, q, r, s, t, u, v are possible?

Since there are seven letters the number of permutations possible is given by $7! = 5\,040$.

However, we are often interested in a slightly more complicated type of permutation problem. How many ways are there to order three letters selected from the alphabet? In other words, how many permutations are possible if we choose three different elements from a set of 26 different elements. It is a lot more difficult to simply write down all the possibilities, but we can figure out how many possibilities there are using the same method as before. Since we are choosing three elements we need three spaces, one for each element. In the first space there are 26 possible elements we could put. In the second space we can put one of the 25 remaining elements. And in the third space we can put one of the 24 elements that are still remaining,

$$\underbrace{}_{\substack{26 \\ \text{possibilites}}} \quad \underbrace{}_{\substack{25 \\ \text{possibilites}}} \quad \underbrace{}_{\substack{24 \\ \text{possibility}}}$$

so we have $26 \cdot 25 \cdot 24 = 15\,600$ possible orderings. We can come up with a nice formula for this,

$$26 \cdot 25 \cdot 24 = \frac{26 \cdot 25 \cdot 24 \cdot \cancel{23} \cdot \cancel{22} \cdot \cancel{21} \cdots \cancel{3} \cdot \cancel{2} \cdot \cancel{1}}{\cancel{23} \cdot \cancel{22} \cdot \cancel{21} \cdots \cancel{3} \cdot \cancel{2} \cdot \cancel{1}}$$

$$= \frac{26!}{23!}$$

$$= \frac{26!}{(26-3)!}.$$

In general we may wish to choose r objects from a set of n objects to fill r spaces where the ordering of the objects matters. We have the number of permutations of r objects selected from a set of n objects, where $n > r$, given by

$$\frac{n!}{(n-r)!}.$$

The standard notation used for the number of permutations of r objects selected from n objects is P_r^n, which is read out loud as "n-pee-r." Thus we have the formula

$$P_r^n = \frac{n!}{(n-r)!}.$$

Example 7.11

How many different ways are there to arrange 5 letters from the alphabet?

The number of permutations of 5 letters from the 26 letters of the alphabet is given by

$$P_5^{26} = \frac{26!}{(26-5)!} = \frac{26!}{21!} = 7\,893\,600.$$

Example 7.12

You are given a list of 20 different flavors of ice cream and asked to rank your four favorite flavors. How many different possible responses are there?

You are basically being asked to find the number of permutations of four items from 20 items, or P_4^{20}, which is given by

$$P_4^{20} = \frac{20!}{(20-4)!} = \frac{20!}{16!} = 116\,280.$$

Example 7.13

How many different ways are there to select a committee consisting of a chairperson, a secretary, and a treasurer from a group of twelve people?

Here order matters. We may say that the first person selected becomes the chairperson, the second person becomes the secretary, and the third person becomes the treasurer. Thus this becomes a permutation question, how many permutations of three items from 12 items. The answer is given by

$$P_3^{12} = \frac{12!}{(12-3)!} = \frac{12!}{9!} = 1\,320.$$

7.4   COMBINATIONS AND SUBSETS

Now suppose we want to know how many subsets of size r there are in a set of n elements. This is almost like the permutation question, only we no longer care about the order. We only care about the actual **combination** of r elements, not in the orderings, or permutations, of those r elements.

Consider the example where we wanted to know how many permutations of three letters there are from the alphabet. Recall that we found the number of possible orderings by considering how we would fill the three spaces,

$$\underbrace{\quad}_{\substack{26 \\ \text{possibilites}}} \quad \underbrace{\quad}_{\substack{25 \\ \text{possibilites}}} \quad \underbrace{\quad}_{\substack{24 \\ \text{possibility}}}$$

gave us $26 \cdot 25 \cdot 24 = 15\,600$ possible orderings. But what if the three letters were a, b, and c. Then there are a total of six possible ways these three letters can be ordered,

$$abc, \quad acb, \quad bac, \quad bca, \quad cab, \quad cba.$$

Considering how the three letters could be ordered in the three spaces gave us

$$\underbrace{\quad}_{\substack{3 \\ \text{possibilites}}} \quad \underbrace{\quad}_{\substack{2 \\ \text{possibilites}}} \quad \underbrace{\quad}_{\substack{1 \\ \text{possibility}}}$$

or 3! possible permutations. These six permutations are all considered the same combination of three letters. This is true of any combination of three letters. Thus, in order to find the total number of combinations we have to divide the number of permutations of three letters, which was $P_3^{26} = 15\,600$, by $3! = 6$. This gave us a total of $2\,600$ different combinations.

In the general case where we want to know how many combinations of r objects drawn from n objects there are we need to divide the number of permutations, P_r^n, by $r!$ to give us

$$\frac{P_r^n}{r!} = \frac{n!}{r!(n-r)!}.$$

The standard notation for the number of combinations of r objects chosen from n objects is $\binom{n}{r}$, though sometimes C_r^n is also used.[1] Thus we have

$$\binom{n}{r} = \frac{n!}{r!(n-r)!}.$$

Example 7.14

How many different combinations of five letters from the alphabet are there?

The number of combinations of 5 letters from the 26 letters of the alphabet is given by

$$\binom{26}{5} = \frac{26!}{5!(26-5)!} = \frac{26!}{5!21!} = 65\,780.$$

Example 7.15

How many different committees of three people can be selected out of a total of twelve people?

For people on a committee it does not matter how they are arranged, so we are asking how many combinations of three are there from 12 items, which is given by

$$\binom{12}{3} = \frac{12!}{3!(12-3)!} = \frac{12!}{3!9!} = 220.$$

Example 7.16

How many subsets of size two are there from a set of size n?

We want to find the number of combinations of 2 elements from a set of n elements. This is given by

$$\binom{n}{2} = \frac{n!}{2!(n-2)!} = \frac{n(n-1)}{2} = \frac{n^2 - n}{2} = 0.5n^2 - 0.5n.$$

---

[1] Other notation for P_r^n and C_r^n that one encounters are nPr and nCr or $P(n,r)$ and $C(n,r)$.

7.5 PERMUTATION AND COMBINATION EXAMPLES

So far we have covered the addition principle, the multiplication principle, permutations, and combinations. Most real-world and computer science problems require us to apply more than one of these ideas at the same time and often require a little creativity.

Example 7.17

A student wants to go from her home X to her friend's home Z. To get from X to Z she must travel through Y. From X to Y there are 3 different bus routes or 5 different train routs. To go from Y to Z there are 4 bus routes or 2 train routes. How many different routes are there from X to Z?

We use the addition principle to find the number of ways from X to Y, which is given by $3 + 5 = 8$. We also use the addition principle to find the number of ways from Y to Z, which is given by $4 + 2 = 6$. Then to find the number of ways from X to Z we use the multiplication principle to get $8 \cdot 6 = 48$ different routes.

Example 7.18

An **anagram** of a word is a permutation of the letters of the word. How many anagrams are there for the following words? (a) ANSWER, (b) HELLO, and (c) MISSISSIPPI

(a) The word ANSWER has six different letters so the number of permutations is given by $P_6^6 = 6! = 720$.

(b) The word HELLO has five letters but two of the letters are L. There are $P_5^5 = 5! = 120$ ways of permuting five letters. However, switching the two Ls does not change permutation so there are $120/2 = 60$ different permutations.

(c) The word MISSISSIPPI has 11 letters and there are $P_{11}^{11} = 11!$ different ways of permuting 11 letters. However, there are 4 Ss, 4 Is, and 2 Ps. Switching any of the 4 Ss does not change the anagram, nor does switching any of the 4 Is or the 2 Ps. There are 4! ways to switch the 4 Ss, 4! ways to switch the 4 Is, and 2! ways of switching the 2 Ps. Thus the total number of anagrams is

$$\frac{11!}{4! \cdot 4! \cdot 2!} = 34\,650.$$

Example 7.19

Let $X = \{1, 2, 3, 4, 5\}$ and $Y = \{a, b, c, d, e, f, g\}$. How many different functions $f : X \to Y$ are there?

Recalling the definition of functions, we know that each element in the domain must be sent to an element in the codomain. We use the multiplication principle. Each of the 5 elements in the domain could be sent to one of the 7 elements in the codomain. Thus the number of possible functions is given by

$$7 \times 7 \times 7 \times 7 \times 7 = 7^5 = 16\,807.$$

Example 7.20

Let $X = \{1, 2, 3, 4, 5\}$ and $Y = \{a, b, c, d, e, f, g\}$. How many different one-to-one functions $f : X \to Y$ are there?

Recalling the definition of one-to-one functions, we know that each element in the domain must be sent to a different element in the codomain. In other words, how many permutations are of of 5 elements from 7 elements. This is

$$P_5^7 = \frac{7!}{(7-5)!} = \frac{7!}{2!} = 2\,520.$$

Example 7.21

A committee consisting of three women and three men are to be chosen from a group of 22 women and 27 men. In how many ways can this be done?

The number of ways we can choose three women from 22 women is given by $\binom{22}{3}$ and the number of ways we can choose three men from 27 men is given by $\binom{27}{3}$. We then use the multiplication principle to find the total number of ways the committee can be formed,

$$\binom{22}{3} \cdot \binom{27}{3} = 4\,504\,500.$$

Example 7.22

A group of 20 individuals has to elect an executive board consisting of a president, a secretary, and a treasurer and three at-large trustees. how many different possible ways are there to do this?

For the first three positions (president, secretary, treasurer) order matters, but for the second three positions (three at-large trustees) order does not matter. If we choose the first three positions first there are P_3^{20} possibilities. We then choose the second three positions where there are now C_3^{17} possibilities. We now use the multiplication principle to get the total number of possibilities.

$$P_3^{20} \cdot C_3^{17} = (6\,840)(680) = 4\,651\,200.$$

But suppose we decided to first choose the three at-large trustees and then choose the president, secretary, and treasurer. In this case the number of possibilities for the first three positions is C_3^{20} and the number of possibility for the second three positions is P_3^{17}. Using the multiplication principle we have

$$C_3^{20} \cdot P_3^{17} = (1\,140)(4\,080) = 4\,651\,200.$$

It should not surprise you that these numbers are the same.

Example 7.23

What is the maximum number of times the **for-do** loop in step 2 is executed? What is the maximum number of times the **for-do** loop in 2.1 is executed? What is the maximum number of times the **if-then** conditional control in setp 2.1.1 is executed?

> Algorithm: Checks if a string of integers $x_1, x_2, \ldots, x_n$ contains any integer more than once.
>
> 1. Input string $x_1, \ldots, x_n$.
> 2. **For** $i = 1$ to $n - 1$ **do**
>   2.1 **For** $j = i + 1$ to n **do**
>   2.1.1 **If** $x_i = x_j$ **then**
>   2.1.1.1 Return "There is duplicate integer in the string."
> 3. Return "There are no duplicate integers in the string."

This algorithm is a little more complicated than the algorithms in section 7.2. We begin with the **for-do** loop in step 2. This **for-do** loop runs from $i = 1$ to $n - 1$ so it is executed a maximum of $n - 1$ times. However, the **for-do** loop in step 2.1 runs from $j = 1 + 1$ to n. So, when $i = 1$ this loop runs from $j = 2$ to n, or $n - 1$ times. When $i = 2$ this loop runs from $j = 3$ to n, or $n - 2$ times, and so on.

A better way to think about this algorithm it so recognize that it does one comparison (in step 2.1.1) for each set of two non-equal indices i and j. In other words, once for each set $\{i, j\}$ of two numbers in the set $\{1, 2, \ldots, n\}$. We can use combinations to find this,

$$f(n) = \binom{n}{2} = \frac{n(n-1)}{2} = \frac{n^2 - n}{2} = 0.5n^2 - 0.5n,$$

so the maximum number of times the **for-do** loop in step 2.1 is executed is $0.5n^2 - 0.5n$ times. The maximum number of times the **if-then** loop in step 2.1.1 is executed is the same.

7.6 PROBLEMS

Question 7.1 *How many five digit decimal number are there that contain only odd digits?*

Question 7.2 *How many five digit octal number are there that contain only odd digits?*

Question 7.3 *How many five digit hexadecimal number are there that contain only odd digits?*

Question 7.4 *How many different ways can a first, second, and third prize be awarded in a class of 32 students.*

Question 7.5 *How many different ways are there to choose a five person committee that has the positions chair, vice-chair, communications director, secretary, and treasurer from a group of 20 people?*

Question 7.6 *How many different ways are there to choose a committee of five people from a group of 20 people?*

Question 7.7 *An individual needs to travel from X to Z. To get from X to Z one must travel through Y. From X to Y there are 8 different bus routes or 6 different train routs. To go from Y to Z there are 5 bus routes or 9 train routes. How many different routes are there from X to Z?*

Question 7.8 *A school assigns two kinds of identification numbers to its staff. Teachers have an ID that starts with a T which is followed by four digits. Administrators have an ID that starts with an A and is followed by three digits. How many total identification numbers are available for the school to use?*

Question 7.9 *How many anagrams are there for the following words?*

  (a) COMBINE   *(b) PERMUTE*   *(c) ASSIGNMENT*

Question 7.10 *Let $X = \{u, v, w, x, y, z\}$ and $Y = \{1, 2, 3, 4, 5, 6, 7, 8, 9\}$. How many different functions $f : X \to Y$ are there? How many one-to-one functions are there?*

Question 7.11 *Let $X = \{\alpha, \beta, \gamma, \delta\}$ and $Y = \{A, B, C, D, E, F, G, H, I, J, K\}$. How many different functions $f : X \to Y$ are there? How many one-to-one functions are there?*

Question 7.12 *An online quiz has four pools of questions. The first two pools have 10 questions each and the last two pools have 20 questions each. If two questions are drawn at random from each of the first two pools and three questions are drawn at random from each of the second two pools (for a total of ten questions) how many possible quizzes can be generated. (Assume the order of the questions does not matter.)*

Question 7.13 *What is the maximum number of possible times the **while-do** loop in line 4 could execute?*

> 1. Input $x_1, x_2, \ldots, x_n$.
> 2. $i \leftarrow 1$
> 3. *order* $\leftarrow$ *true*
> 4. **While** $i < n$ and *order* = true **do**
>   4.1. **If** $x_i > x_{i+1}$ **then**
>   4.1.1. *order* $\leftarrow$ false
>   4.2. $i \leftarrow i + 1$
> 5. **If** *order* = true **then**
>   5.1. Output "Numbers are in order."
>   **else**
>   5.2. Output "Numbers are out of order."

Question 7.14 *Given two $n \times n$ matrices A and B the following algorithm performs matrix operation. (An $n \times n$ matrix A is an $n \times n$ array of numbers labeled a_{ij} where $1 \leq i \leq n$ and $1 \leq j \leq n$. Thus the matrix A consists of n^2 numbers. The matrix B is defined similarly.) What is the maximum number of times the **for-do** loop in step 2 could execute? What is the maximum number of times the **for-do** loop in step 2.1 could execute? What is the maximum number of times the **for-do** loop in step 2.1.2 could execute?*

1. Input a_{ij} and b_{ij} for $1 \leq i \leq n$ and $1 \leq j \leq n$.
2. **For** $i = 1$ to n **do**
   2.1. **For** $j = 1$ to n **do**
      2.1.1. *sum* $\leftarrow 0$
      2.1.2. **For** $k = i$ to n **do**
         2.1.2.1 *sum* $\leftarrow sum + x_{ik} \times x_{kj}$
      2.1.3 $c_{ij} \leftarrow sum$
3. Output c_{ij}.

Question 7.15 *The following is an algorithm that sorts a list of n numbers into increasing order. What is the maximum number of times the **for-do** loop in step 2 could execute? What is the maximum number of times the **while-do** loop in step 2.3 could execute?*

1. Input $x_1, x_2, \ldots, x_n$.
2. **For** $i = 2$ to n **do**
   2.1. *insert* $\leftarrow x_i$
   2.2. $j \leftarrow i - 1$
   2.3. **While** $j \geq 1$ and $x_j > insert$ **do**
      2.3.1. $x_{j+1} \leftarrow x_j$
      2.3.2. $j \leftarrow j - 1$
   2.4. $x_j \leftarrow insert$
3. Output $x_1, x_2, \ldots, x_n$.

Question 7.16 *The following is Warshall's algorithm to determine if there exists a directed path from v_i to v_j. What is the maximum number of times the **for-do** loop in step 2 is executed? What is the maximum number of times the **for-do** loop in step 3 is executed? What is the maximum number of times the **for-do** loop in step 3.1 is executed? What is the maximum number of times the **if-then** conditional control in step 3.1.1 is executed? What is the maximum number of times the **for-do** loop in step 3.1.1.1 is executed?*

1. Input adjacency matrix (a_{ij}).
2. **For** $i = 1$ to n **do**
   2.1. $a_{ii} = 1$
3. **For** $i = 1$ to n **do**
   3.1 **For** $j = 1$ to n **do**
      3.1.1 **If** $a_{ij} = 1$ **then**
         3.1.1.1 **For** $k = 1$ to n **do**
            3.1.1.1.1 $a_{jk} \leftarrow a_{jk} + a_{ik}$ (Boolean Addition)
4. Output (a_{ij}).

CHAPTER 8

Algorithmic Complexity

One of the most important jobs for programmers is choosing, or designing, the fastest algorithms to use when writing programs. Therefore, it is extremely important that they have some way to compare the speed of different algorithms. But algorithms can be very complicated and very different from each other, so comparing them is not easy. This chapter focuses on how this is done.

8.1   OVERVIEW OF ALGORITHMIC COMPLEXITY

How long does it take for an algorithm, implemented as a computer program, to run? This is an extremely important question for computer scientists and programmers. The answer depends on many different factors, like how powerful the computer is, the programming language, the computer operating system, the computer memory, and so on. But it also depends on the algorithm being used and what the input is. In this chapter we will only consider two factors, the algorithm and the input, and ignore the other factors. We are basically analyzing how complex the algorithm is.

Because this chapter is quite complicated we will begin with a big-picture overview of algorithmic complexity. The following flowchart gives the general idea. Hopefully it will help you to understand all that is going on in this chapter.

Algorithm
↓
Find time-complexity function for algorithm.
↓
Find big-$\mathcal{O}$ category for time-complexity function.
↓
Use big-$\mathcal{O}$ category to rank algorithms by speed.

The general idea is that given any algorithm one can find a function associated with that algorithm that essentially describes how long it takes the algorithm to run. Because algorithms can be so different and so complicated we need to make three simplifying approximations. The function found is called the algorithm's time-complexity function. Once the time-complexity function for the algorithm is obtained we find what is called the big-$\mathcal{O}$ (pronounced "big-oh") category for the time-complexity function. This requires us to make another two simplifying approximations. Doing this essentially groups together all the algorithms that have roughly the same speed into the same category. In other words, we usually think of all the algorithms whose time-complexity functions are in the same big-$\mathcal{O}$ category as being equally fast. Finally, we then use the big-$\mathcal{O}$ categories to rank algorithms by speed.

As we have already stated, algorithmic complexity analysis requires us to make a total of five simplifying approximations. These five approximations are listed below. The first three approximations are needed to find the time-complexity function for an algorithm and are studied in section 8.2. The last two approximations are related to finding the big-$\mathcal{O}$ category of the time-complexity function and will be studied in section 8.4.

Approximation One: Count only the most frequently executed operations.

Approximation Two: Of the most frequently executed operations, count only the most time-consuming operation.

Approximation Three: Analyze the worst-case scenario.

Approximation Four: Assume the input is large.

Approximation Five: Do not distinguish between two time-complexities that are a constant multiple of each other.

8.2 TIME-COMPLEXITY FUNCTIONS

We begin this section by giving the definition of the time-complexity function for an algorithm. But it will take the rest of the section to fully understand this definition.

Definition 8.1 *The **time-complexity function** of an algorithm is a function*

$$f : \mathbb{N} \longrightarrow \mathbb{N}$$

where $\mathbb{N}$ is the set of natural numbers and where

$$f(n) = \begin{array}{l} \textit{Maximum number of} \\ \textit{dominant operations} \\ \textit{performed by the} \\ \textit{algorithm if the input} \\ \textit{is of size n.} \end{array}$$

This definition of time-complexity actually uses the first three approximations. The phrase "dominant operations" means we are using the first and second approximations, counting only the most time consuming frequent operation. The phrase "maximum number" means we are using the third approximations, using only the worst-case scenario. What exactly is meant by the phrase "the input is of size n" depends on the exact nature of the algorithm. In section 8.3 we will look at examples of algorithms and find the time-complexity functions for these algorithms.

What we are doing is using the time-complexity function as a way of indirectly measuring how long an algorithm takes to run. Of course, the answer is not given in terms of a time like seconds or hours or weeks, it is given in terms of how many times the dominant operations are performed in the worst case scenario. How long it takes to run the algorithm depends on the speed of the computer you are using, and every year the speed of computers keeps getting faster, but the relative times needed to perform operations stays the same.

Approximations One and Two

The first two approximations are related to finding the dominant operation in the algorithm. The operations performed in the algorithms in this book are assignments ($\longleftarrow$), comparisons ($=, \neq, >, <, \not<, \not>$), addition, subtraction, multiplications, and division. According to the first approximation, we are interested in the operations that are executed most frequently in the algorithm. Consider the following algorithm.

Algorithm: Calculates x^n.

1. Input x and n.
2. $answer \longleftarrow x$
3. **For** $i = 2$ to n **do**
    3.1. $answer \longleftarrow answer \times x$
4. Output $answer$.

In step 2 there is an assignment and in step 3.1 there is both an assignment and a multiplication. The assignment in step 2 is executed only onec, whereas the assignment and the multiplication in step 3.1 are executed each time the **for-do** loop is executed. The first approximation tells us that we are only interested in the operations that are frequently executed, therefore we are only interested in the operations in step 3.1, not in the operation in step 2 since it is only executed once. In practice this means we are mostly interested in the operations that happen inside loops.

The second approximation states that of the most frequently executed operations, we only count the most time-consuming operation. This means we only count the slowest operation. The most time-consuming frequent operation is the **dominant operation** of the algorithm. Here we list the operations from fastest to slowest:

- Assignment ($\longleftarrow$)
- Comparisons ($=, \neq, >, <, \nless, \ngtr$)
- Addition and subtraction ($+, -$)
- Multiplication and division($\times, /$)

We would like to know how long it would a computer to perform each of these operations. In general we cannot know for sure, but we can say that assignments are usually faster than comparisons, which are usually faster than addition and subtraction, which in turn are usually faster than multiplication and division.

Let us consider the two operations from step 3.1 of the above algorithm. The two operations were assignment and multiplication. According to the above list, assignment is much faster than multiplication, so multiplication is the most time-consuming frequent operation. That means that multiplication is the dominant operation in this algorithm. Let us look at another example.

Algorithm: Finds the smallest value in a list of numbers.

1. Input the number of values in list, n.
2. Input the list of numbers $x_1, x_2, \ldots, x_n$.
3. $min \longleftarrow x_1$
4. **For** $i = 2$ to n **do**
    4.1 **If** $x_i < min$ **then**
        4.1.1 $min \longleftarrow x_i$
5. Output min.

There is an assignment $\longleftarrow$ in step 3, which is outside the **for-do** loop. Inside the **for-do** loop there are two operations, a comparison $<$ in step 4.1 and an assignment $\longleftarrow$ in step 4.1.1. Using approximation one we can ignore the assignment in step 3 because it is only executed once. We are only interested in the operations inside the **for-do** loop. Then, using approximation two we only consider the most time-consuming of these two operations. Since comparisons are slower than assignments then the comparison in step 4.1 is the dominant operation in the algorithm.

Figuring out the dominant operation is a little difficult sometimes. Notice how we said that assignments are usually faster than comparisons, which are usually faster than addition and subtraction, which in turn are usually faster than multiplication and division. If we are only adding one, or subtracting one, or multiplying by one, we generally consider these as very fast operations. It is easy to simply add or subtract one, and multiplying by one simply gives us the original number back. So if an operation is only adding or subtracting one or multiplying by one then it is usually not considered a dominant operation. Look at the below algorithm.

> Algorithm: Sorts a list $x_1, x_2, \ldots, x_n$ of n numbers into increasing order.
>
> 1. Input $x_1, x_2, \ldots, x_n$.
> 2. **For** $i = 2$ to n **do**
>   2.1. *insert* $\longleftarrow x_i$
>   2.2. $j \longleftarrow i - 1$
>   2.3. **While** $j \geq 1$ and $x_j > insert$ **do**
>   2.3.1. $x_{j+1} \longleftarrow x_j$
>   2.3.2. $j \longleftarrow j - 1$
>   2.4. $x_j \longleftarrow insert$
> 3. Output $x_1, x_2, \ldots, x_n$.

There are assignments in steps 2.1, 2.2, 2.3.1, 2.3.2, and 2.4. Assignments are all quite fast. There are two comparisons in step 2.3. And it appears there is an addition in step 2.3.1 and subtractions in steps 2.2 and 2.3.2. Using the above list we might be tempted to say that the addition and subtractions are the dominant operations. But these are simply adding or subtracting one. The comparison $x_j > insert$ is almost certainly more time consuming. Therefore in this case the comparisons would be the dominant operation. This is the sort of thing that will be more obvious to you after you have taken some programming courses.

Approximation Three

The third approximation states that we should only analyze the worst-case scenario. Consider the following algorithm.

> Algorithm: Searches a string of integers $x_1, \ldots, x_n$ to see if it contains integer s.
>
> 1. Input string $x_1, \ldots, x_n$, n and s.
> 2. **For** $i = 1$ to n **do**
>   2.1. **If** $x_i = s$ **then**
>   2.1.1. Return "String contains s."
> 3. Return "String does not contain s."

Suppose we wanted to use this algorithm to check if integer 7 was contained in each of these two 50 digit strings,

01234567890123456789012345678901234567890123456789,

00000000000000000000000000000000000000000000000000.

For the first string the **for-do** loop is executed eight times. For the second string the **for-do** loop is executed all 50 times. The number of times the **for-do** loop is executed depends on the input. If we are checking to see if integer s is contained in an n digit string then the maximum possible number of times the **for-do** loop could be executed is n times. It may be executed fewer times than that, but n times is the worst-case scenario. When analyzing algorithms we always assume any loops are executed the maximum possible number of times.

8.3   FINDING TIME-COMPLEXITY FUNCTIONS

So far we have discussed the three approximations we need to make in order to find the time-complexity function of an algorithm. In this section we use these approximations on some specific examples of algorithms to figure out their time-complexity functions.

Example 8.1

What is the time-complexity function of this algorithm?

> Algorithm: Searches a string of integers $x_1, \ldots, x_n$ to see if it contains the integer s.
>
> 1. Input string $x_1, \ldots, x_n$, n and s.
> 2. **For** $i = 1$ to n **do**
>   2.1. **If** $x_i = s$ **then**
>   2.1.1. Return "String contains s."
> 3. Return "String does not contain s."

First we ask ourselves, what is the dominant operation in this algorithm. The only operation that is done in this algorithm is checking to see if x_i is equal to s in step 2.1, therefore this comparison is the dominant operation. Next we ask ourselves how many times this operation is performed. We are concerned with the worst-case scenario. If there is an s in the string then "String contains s." is returned in step 2.1.1 and the algorithm ends. But if there is no integer s in the string $x_1, \ldots, x_n$ then the **for-do** loop must run n times, once for each x_i in the string. This is the worst-case scenario. This means the dominant operation is done n times. Thus the time-complexity function is
$$f(n) = n.$$

Example 8.2

What is the time-complexity function of this algorithm?

> Algorithm: Searches two strings of integers $x_1, \ldots, x_n$ and $y_1, \ldots, y_n$ of equal length to see if either string contains integer s.
>
> 1. Input strings $x_1, \ldots, x_n$ and $y_1, \ldots, y_n$, n and s.
> 2. **For** $i = 1$ to n **do**
>   2.1. **If** $x_i = s$ **then**
>   2.1.1. Return "String one contains integer s."
> 3. **For** $i = 1$ to n **do**
>   3.1. **If** $y_i = s$ **then**
>   3.1.1. Return "String two contains integer s."
> 4. Return "Neither string contains integer s."

The analysis of this algorithm is almost exactly the same as in the last example. In the worst-case scenario the integer s is not contained in either string. This would mean the **for-do** loop in step 2 is carried out n times and then the **for-do** loop in step 3 is also carried out n times. The addition principle gives the time-complexity function as
$$f(n) = n + n = 2n.$$

Example 8.3

What is the time-complexity function of this algorithm?

> Algorithm: Checks if two strings $x_1, \ldots, x_n$ and $y_1, \ldots, y_n$ contain a common integer. (That is, check and see if there is an integer s that is contained in both strings.)
>
> 1. Input strings $x_1, \ldots, x_n$ and $y_1, \ldots, y_n$.
> 2. **For** $i = 1$ to n **do**
>   2.1 **For** $j = 1$ to n **do**
>   2.1.1 **If** $x_i = y_j$ **then**
>   2.1.1.1 Return "There is an element common to both strings."
> 3. Return "There are no elements in common to both strings."

The dominant operation in this algorithm is the comparison in step 2.1.1 where we check to see if x_i is equal to y_j. How many times is the operation executed in the worst-case scenario? Notice that this operation is contained in a nested loop. There is the **for-do** loop of step 2 and then the **for-do** loop of step 2.1. The worst-case scenario happens if there is no integer that is common to both strings. If this is the case then for $i = 1$ the **for-do** loop in step 2.1 runs n times. Then for $i = 2$ it runs n times again. Then for $i = 3$ it runs n times again, and so on. The multiplication principle tells us the dominant operation executes a total of n^2 times. This means the time-complexity function for this algorithm is
$$f(n) = n^2.$$

Example 8.4

What is the time-complexity function of this algorithm?

> Algorithm: Checks to see if a string of integers $x_1, x_2, \ldots, x_n$ contains any integer more than once.
>
> 1. Input string $x_1, \ldots, x_n$.
> 2. **For** $i = 1$ to $n - 1$ **do**
>   2.1 **For** $j = i + 1$ to n **do**
>   2.1.1 **If** $x_i = x_j$ **then**
>   2.1.1.1 Return "There is duplicate integer in the string."
> 3. Return "There are no duplicate integers in the string."

This algorithm is similar to the last one, but with one notable difference. The dominant operation is the comparison in step 2.1.1, which is contained in a double loop. The first **for-do** runs from $i = 1$ to $n - 1$ but the second for-do loop only runs from $j = i + 1$ to n. Roughly, this algorithm does half the work of the last example so $f(n) \approx \frac{n^2}{2}$. Another way to think about it is that the algorithm does one comparison for each set of two non-equal indices i and j. This is the same as the number of sets of two different numbers $\{i, j\}$ that are in the set $\{1, 2, \ldots, n\}$. This is a combination problem. Thus the time-complexity function is given by
$$f(n) = \binom{n}{2} = \frac{n(n-1)}{2} = \frac{n^2 - n}{2} = 0.5n^2 - 0.5n.$$

Example 8.5

What is the time-complexity function of this algorithm?

Algorithm: Sorts a list $x_1, x_2, \ldots, x_n$ of n numbers into increasing order.

1. Input $x_1, x_2, \ldots, x_n$.
2. **For** $i = 2$ to n **do**
   2.1. *insert* $\longleftarrow x_i$
   2.2. $j \longleftarrow i - 1$
   2.3. **While** $j \geq 1$ and $x_j > insert$ **do**
      2.3.1. $x_{j+1} \longleftarrow x_j$
      2.3.2. $j \longleftarrow j - 1$
   2.4. $x_j \longleftarrow insert$
3. Output $x_1, x_2, \ldots, x_n$.

We have already determined that the dominant operations in this algorithm are the comparisons in step 2.3. But how many times are these operations performed? First, we have the loop in step 2 happening $n - 1$ times. Each time this loop is executed the variable j is initialized as $i - 1$. Then each time the **while-do** loop is executed the j value is decreased by one. The **while-do** loop continues until $j = 1$. In other words, for each value of i the values of j range form 1 to $i - 1$. This is just like the previous example, the number of times the while-do loop is executed is the same as the number of subsets of two non-equal numbers $\{i, j\}$ that are in the set $\{1, 2, \ldots, n\}$. This is a combination problem. Since there are two comparisons in step 2.3 the time-complexity function is

$$f(n) = 2\binom{n}{2} = 2\left(\frac{n(n-1)}{2}\right) = n^2 - n.$$

Example 8.6

What is the time complexity function for the first algorithm to find x^n.

Algorithm: First algorithm to find x^n.

1. Input x and n.
2. *answer* $\longleftarrow x$
3. **For** $i = 2$ to n **do**
   3.1. *answer* $\longleftarrow$ *answer* $\times x$
4. Output *answer*.

Multiplication is the dominant operation in this algorithm. The dominant operation is executed each time the **for-do** loop is executed, which happens from $i = 2$ to n, or $n - 1$ times. This gives a time complexity function of

$$f_1(n) = n - 1.$$

Example 8.7

What is the time complexity function for the second algorithm to find x^n.

> Algorithm: Second algorithm to find x^n.
>
> 1. Input x and n.
> 2. $d \longleftarrow$ number of bits in n. ($n_{10} = b_d b_{d-1} b_{d-2} \ldots b_{1\,2}$)
> 3. **If** $b_1 = 1$ **then**
>   3.1 *answer* $\longleftarrow x$
>   **else**
>   3.2 *answer* $\longleftarrow 1$
> 4. **For** $i = 2$ to d **do**
>   4.1 $x \longleftarrow x \times x$
>   4.2 **If** $b_i = 1$ **then**
>   4.2.1 *answer* $\longleftarrow$ *answer* $\times x$
> 5. Output *answer*.

We will take a closer look at this algorithm in the next section, but for now notice that the dominant operation in this algorithm is multiplication in steps 4.1 and 4.2.1. What controls the number of times the dominant operation is executed? The number of times the multiplication in step 4.1 happens depends on d, the number of bits of n. The multiplication in step 4.2.1 depends on whether a particular digit b_i in the binary expansion of n is a 1 or a 0. Thus, the number of dominant operations depends entirely on the binary expansion of n.

The **for-do** loop says to repeat steps 4.1 and 4.2 from $i = 2$ to d, or $d-1$ times. The **if-then** conditional control tells us to do the multiplication in step 4.2.1 only if the digit b_i of the decimal representation of n is 1. Here we apply approximation three and assume the worst-case scenario, that every digit in the decimal representation of n is 1. Hence this multiplication is done $d-1$ times as well. Thus, each time the for-do loop happens then in the worst case scenario there are two multiplications, which means a total of $2(d-1)$ multiplications happen. As will be explained in the next section, $d = \lfloor \log_2(n) \rfloor + 1$, so we have

$$2(d-1) = 2\lfloor \log_2(n) \rfloor \leq 2\log_2(n)$$

so the time-complexity function for this second algorithm is given by

$$f_2(n) = 2\log_2(n).$$

A word of caution, by replacing $\lfloor \log_2(n) \rfloor$ by $\log_2(n)$ the codomain of this time-complexity function is no longer $\mathbb{N}$ but $\mathbb{R}$, but this is something that is done quite often in practice.

8.4 BIG-O NOTATION

What we are now interested in is a way to compare algorithms with different time-complexity functions. In computer science the big-$\mathcal{O}$ notation is used to help classify how fast different algorithms are. We first explain the two approximations that are used in determining the big-$\mathcal{O}$ category for an algorithm before introducing the definition toward the end of the section.

Approximation Four

In the fourth approximation we assume the input is large. In order to understand why approximation four is important we will consider the two different algorithms for computing x^n. This is the first algorithm.

Algorithm: First algorithm to find x^n.

1. Input x and n.
2. *answer* ⟵ x
3. **For** $i = 2$ to n **do**
   3.1. *answer* ⟵ *answer* $\times x$
4. Output *answer*.

And here is the second algorithm for computing x^n. Notice in the second step we are asked to find the number of bits of n. The word **bits** comes from **binary digits** and simply means the number of digits that is in the binary number equal to n. Thus, if n is written as a binary number with digits $b_d, \ldots, b_1$ then n is d bits. For example, if $n = 41_{10} = 101001_2$ then $d = 6$ and $b_6 = 1$, $b_5 = 0$, $b_4 = 1$, $b_3 = 0$, $b_2 = 0$, and $b_1 = 1$. Thus, n has six bits. In general, the number of bits (binary digits) contained in the decimal number n is given by the formula

$$d = \lfloor \log_2(n) \rfloor + 1.$$

Algorithm: Second algorithm to find x^n.

1. Input x and n.
2. d ⟵ number of bits in n. $(n_{10} = b_d b_{d-1} b_{d-2} \ldots b_1{}_2)$
3. **If** $b_1 = 1$ **then**
   3.1 *answer* ⟵ x
   else
   3.2 *answer* ⟵ 1
4. **For** $i = 2$ to d **do**
   4.1 x ⟵ $x \times x$
   4.2 **If** $b_i = 1$ **then**
      4.2.1 *answer* ⟵ *answer* $\times x$
5. Output *answer*.

The way the first algorithm works for finding x^n is obvious, but the way the second algorithm works for finding x^n may not be obvious at all. We will use a simple example to illustrate how these algorithms differ. Suppose you wanted to find 2^{64}. Using the first algorithm would require 63 multiplications since

$$7^{64} = \underbrace{7 \times 7 \times 7 \times 7 \times 7 \times 7 \times 7 \times 7 \times \cdots \times 7}_{64 \text{ sevens and 63 multiplications}}.$$

But now notice that

$$7^{64} = 7^{32} \times 7^{16} \times 7^8 \times 7^4 \times 7^2 \times 7 \times 7.$$

If we are clever then we can drastically reduce the number of multiplications necessary to find 7^{64}. First we find 7×7 which takes one multiplication and that gives us 7^2. Then we do $7^2 \times 7^2$ which is a second multiplication and gives us 7^4. Then we do $7^4 \times 7^4$ which is a third multiplication and gives us 7^8. Then $7^8 \times 7^8$ is a fourth multiplication and gives us

7^{16}. Then $7^{16} \times 7^{16}$ is a fifth multiplication and gives us 7^{32}. Finally $7^{32} \times 7^{32}$ is a sixth multiplication which gives us 7^{64}. Thus by being clever we have reduced 63 multiplications to six multiplications. This is a huge improvement.

Essentially the second algorithm does just this, it attempts to reduce the number of multiplications necessary to find x^n. A carefully constructed algorithm can reduce the amount of work necessary. This is where the fourth approximation becomes important.

Example 8.8

Suppose we wanted to find 7^5. How many multiplications would be needed in each of the two algorithms to find x^n?

This is easily seen using the time complexity functions for these algorithms that were found in the last section. The time complexity function for the first algorithm is $f_1(n) = n - 1$, so for the first algorithm we have

$$f_1(5) = 5 - 1 = 4.$$

The time complexity function for the second algorithm is $f_2(n) = 2 \log_2(n)$, so for the second algorithm we have

$$f_2(4) = 2 \log_2(5) \approx 4.64.$$

Based on this we might easily conclude that the first algorithm is actually a little better than the second algorithm.

Example 8.9

Suppose we wanted to find 7^{1024}. How many multiplications would be needed in each of the two algorithms to find x^n?

This is easily seen using the time complexity functions for these algorithms that were found in the last section. For the first algorithm we have

$$f_1(1024) = 1024 - 1 = 1023$$

and for the second algorithm we have

$$f_2(1024) = 2 \log_2(1024) = 20.$$

When we assume the input is large, it is very clear that the second algorithm is much faster than the first algorithm.

Here are a few more examples that make it clear why it is important to assume that the input n is large.

Example 8.10

Suppose the time-complexity of algorithm one is $f_1(n) = n^2$ and the time-complexity of the second algorithm is $f_2(n) = n^3$. Which algorithm is faster?

It should be clear that for any natural number $n > 1$ that n^2 is a smaller number than n^3. That means that algorithm one performs fewer dominant operations than algorithm two and so it is faster. You can see this using a table.

n	$f_1(n) = n^2$	$f_2(n) = n^3$
1	1	1
5	25	125
10	100	1,000
100	10,000	1,000,000

Example 8.11

Now suppose the time-complexity of algorithm one is $f_1(n) = 1,000,000n^2 = 10^6 n^2$ and the time-complexity of the second algorithm is $f_2(n) = n^3$. Which algorithm is faster?

Let us use the same values of n that we used in the last example.

n	$f_1(n) = 1,000,000n^2$	$f_2(n) = n^3$
1	1,000,000	1
5	25,000,000	125
10	100,000,000	1,000
100	10,000,000,000	1,000,000

According to this table, for all the values of n that we have chosen, algorithm two does fewer dominant operations than algorithm one, thus we might think that algorithm two is faster. This is where assumption four becomes very important. Let us make another table for much larger values of n.

n	$f_1(n) = 10^6 n^2$	$f_2(n) = n^3$
1	10^6	1
10^6	10^{18}	10^{18}
10^{15}	10^{36}	10^{45}
10^{20}	10^{46}	10^{60}

Up until $n = 10^6$ algorithm two is faster, but when n is larger than 10^6 algorithm one is faster. So, for *large* n, algorithm one is much faster.

Approximation Five

In the fifth approximation we decide not to distinguish between two time-complexities that are a constant multiple of each other. A function f_2 is said to be a **constant multiple** of f_1 if there is some real number c such that $f_2(n) = cf_1(n)$.

Example 8.12

Determine if two given time-complexity functions are a constant multiple of each other.

1. Are $f_1(n) = 5n^2$ and $f_2(n) = 50n^2$ constant multiples of each other? Since

$$f_2(n) = 10f_1(n)$$

then f_1 and f_2 are constant multiples of each other. Here the constant is 10.

2. Are $f_1(n) = 500n^4$ and $f_2(n) = \frac{n^4}{100}$ constant multiples of each other? Since

$$f_1(n) = 50,000 f_2(n)$$

then f_1 and f_2 are constant multiples of each other. Here the constant is $50,000$. We also have

$$f_2(n) = \frac{1}{50,000} f_1(n).$$

Here the constant is $\frac{1}{50,000}$. So, it does not matter if we write f_1 in terms of f_2 or f_2 in terms of f_1.

3. Are $f_1(n) = 200n^3$ and $f_2(n) = 600n^4$ constant multiples of each other? We have

$$f_2(n) = 3n f_1(n).$$

Here the multiple is $3n$. Since n is a variable then $3n$ is not a constant but changes when the value of n changes and so f_2 is not a constant multiple of f_1.

How this is applied is through the following definition.

Definition 8.2 *Suppose we have two time-complexity functions f and g. We say that f is $\mathcal{O}(g)$ if there is some positive number c such that*

$$f(n) \leq cg(n)$$

for all n that are very large. $\mathcal{O}(g)$ is read "big-oh of g."

Example 8.13

Suppose $f(n) = 5n + 20$ and $g(n) = n$. Show that f is $\mathcal{O}(g)$.

In order to show that f is $\mathcal{O}(g)$ we need to find some c such that $f(n) \leq cg(n)$ for all very large values of n. If we choose $c = 5$ then it is clear that $5n + 20 \not\leq 5n$ for any values of n no matter how large. But suppose we choose $c = 6$. Then we have

$$5n + 20 \leq 6n.$$

When is this inequality true? By subtracting $5n$ from both sides of the inequality we can see it is true when

$$20 \leq n.$$

Thus, for all values of n larger than or equal to 20 we have $5n + 20 \leq 6n$ and so f is $\mathcal{O}(g)$. We could also say that f is $\mathcal{O}(n)$.

Example 8.14

Suppose that $f(n) = 10n^5 - 7n^4 + 20n^2 + 100$ and $g(n) = n^5$. Show that f is $\mathcal{O}(g) = \mathcal{O}(n^5)$.

Since f and g are time-complexity functions and $f : \mathbb{N} \longrightarrow \mathbb{N}$ then the domain of both f and g is $\mathbb{N}$. This means that n is a positive number. A little thought should convince you that

$$10n^5 - 7n^4 + 20n^2 + 100 \leq 10n^5 + 20n^2 + 100.$$

A little more thought should convince you that

$$10n^5 + 20n^2 + 100 \leq 10n^5 + 20n^5 + 100n^5.$$

And finally we have

$$10n^5 + 20n^5 + 100n^5 = 130n^5.$$

Putting all of this together we have

$$10n^5 - 7n^4 + 20n^2 + 100 \leq 130n^5.$$

Thus f is $\mathcal{O}(n^5)$. In general, if f is any polynomial of degree k then f is $\mathcal{O}(n^k)$. If f is a sum of different terms, the f is big-$\mathcal{O}$ of the fastest growing term in the sum.

The functions that are of most interest in analyzing algorithms in computer science are 1, $\log_2(n)$, $\sqrt{n}$, n, $n\log_2(n)$, n^2, and 2^n. Fig. 8.1 plots these functions for $0 \leq n \leq 100$. The function $f(n) = 1$ is not much of a function, no matter what the value of the input n is the output is always 1. It should be apparent the function 1 does not grow at all as n increases. The function $\log_2(n)$ grows very slowly, $\sqrt{n}$ grows a little faster, n grows linearly, $n\log_2(n)$ grows faster yet, n^2 grows even faster, and 2^n grows fastest of the functions listed. By looking at Fig. 8.1 we can see that for large n we have the following inequalities,

$$1 < \log_2(n) < \sqrt{n} < n < n\log_2(n) < n^2 < 2^n.$$

Hopefully you also remember that

$$1 < n < n^2 < n^3 < n^4 < n^5 < n^6 < \cdots.$$

Also, for any sufficiently large n we have

$$n^k < 2^n$$

for any given value of $k \in \mathbb{N}$. These hierarchies of functions can be used to help us figure out what big-$\mathcal{O}$ category a time-complexity function is in.

Example 8.15

Consider the time-complexity function $f(n) = 25\sqrt{n} + 100n + 5000$. Find the function g such that f is $\mathcal{O}(g)$.

All we have to do is look at each term in the polynomial and use the above string of inequalities. From the string of inequalities we know that for large n we have $\sqrt{n} < n$

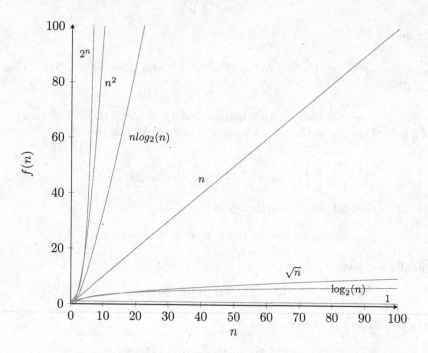

Figure 8.1 A comparison of the growth rates for some common time-complexity functions.

and $1 < n$, which gives us

$$25\sqrt{n} + 100n + 5000 < 25n + 100n + 5000n$$
$$< 5125n,$$

so f is $\mathcal{O}(n)$.

Example 8.16

Consider the time-complexity function $f(n) = 20\log_2(n) + 10\sqrt{n} + 50n\log_2(n)$. Find the function g such that f is $\mathcal{O}(g)$.

Again, we rely on the string of inequalities above. For large n we know that $\log_2(n) < n\log_2(n)$ and $\sqrt{n} < n\log_2(n)$, which gives us

$$20\log_2(n) + 10\sqrt{n} + 50n\log_2(n) < 20n\log_2(n) + 10n\log_2(n) + 50n\log_2(n)$$
$$< 80n\log_2(n),$$

so f is $\mathcal{O}(n\log_2(n))$.

Example 8.17

Find the big-$\mathcal{O}$ category for the time-complexity function $f(n) = 30n\log_2(n) + 50\sqrt{n} + 200n$.

Of course we could continue exactly as in the last two examples, but the key point in the last two examples is that we look for the slowest term in $f(n)$. The first term has an $n\log_2(n)$, the second term has a $\sqrt{n}$, and the third term has an n. Using the chain of inequalities we can see that

$$\sqrt{n} < n < n\log_2(n)$$

thus the slowest term involves $n\log_2(n)$. This is our big-$\mathcal{O}$ category. Thus $f \in \mathcal{O}(n\log_2(n))$.

Example 8.18

Find the big-$\mathcal{O}$ category for the time-complexity function $f(n) = 100\log_2(n) + 2n + 50n^2 + 5000$.

The first term of f has an $\log_2(n)$, the second term has an n, the third term has an n^2 and the fourth term is simply the constant 5000. Since $5000 = 5000(1)$ we could say that this term has a 1. Putting these in order in terms of speed gives us

$$1 < \log_2(n) < n < n^2.$$

The slowest of these is n^2 so $f \in \mathcal{O}(n^2)$.

Now we can actually understand the fifth approximation better. Consider the following four time-complexity functions,

$$f_1(n) = \sqrt{n}, \qquad f_2(n) = 10\sqrt{n}, \qquad f_3(n) = 500\sqrt{n}, \qquad f_4(n) = 10^5\sqrt{n}.$$

These four time-complexity functions are all a constant multiple of each other. Also, all four of these time-complexity functions are $\mathcal{O}(\sqrt{n})$ and so all four of these algorithms are in the same big-$\mathcal{O}$ category, the category of algorithms that have time-complexity functions that are $\mathcal{O}(\sqrt{n})$.

Of course, some algorithms might be a little faster and some might be a little slower, but relatively speaking their speeds are all fairly close together and, for very large n, all are much faster than any algorithm in a slower category like $\mathcal{O}(n)$ or $\mathcal{O}(n\log_2(n))$. Therefore big-$\mathcal{O}$ categories are used to rank algorithm speeds and all algorithms in the same big-$\mathcal{O}$ category are considered to be equally fast. Because all algorithms with time-complexity functions in the same category are considered to be essentially the same this means we do not distinguish between two time-complexities that are a constant multiple of each other. This explains the wording of approximation five very nicely.

8.5   RANKING ALGORITHMS

Not only do we use the above hierarchies of functions to help us decide what big-$\mathcal{O}$ category a time-complexity function is in, but we also use them to help us rank the different big-$\mathcal{O}$

categories themselves. Rewriting the above hierarchies we have, for sufficiently large n,

$$1 < \log_2(n) < \sqrt{n} < n < n\log_2(n) < n^2 < n^3 < n^4 < \cdots < n^k < 2^n$$

where k is any integer.

Thus, algorithms that have time complexity functions that are $\mathcal{O}(1)$ are constant time algorithms. In other words, how long the algorithm takes to execute does not depend on the size of the input n. Algorithms that have time-complexity functions that are $\mathcal{O}(\log_2(n))$ are considered fairly fast; algorithms that have time-complexity functions that are $\mathcal{O}(\sqrt{n})$ are slower, but still fast; and so on. An algorithm that has a time-complexity function which is $\mathcal{O}(2^n)$ is considered very slow.

Example 8.19

Suppose algorithm one has time-complexity function $f_1 \in \mathcal{O}(\log_2(n))$ and algorithm two has time-complexity function $f_2 \in \mathcal{O}(\sqrt{n})$. Which algorithm is faster?

Comparing the big-$\mathcal{O}$ categories we see that

$$\log_2(n) < \sqrt{n}$$

so algorithm one is faster than algorithm two.

Example 8.20

Suppose algorithm one has time-complexity function $f_1 \in \mathcal{O}(n^3)$ and algorithm two has time-complexity function $f_2 \in \mathcal{O}(n)$. Which algorithm is faster?

Comparing the big-$\mathcal{O}$ categories we see that

$$n < n^3$$

so algorithm two is faster than algorithm one. We could also say that algorithm one is slower than algorithm two.

Example 8.21

Suppose algorithm one has time-complexity function $f_1(n) = 50n\log_2(n) + 200\sqrt{n}$ and algorithm two has time-complexity function $f_2(n) = 30\log_2(n) + 500n$. Which algorithm is slower?

First we have to find the big-$\mathcal{O}$ categories of each algorithm. It is easy to see that $f_1 \in \mathcal{O}(n\log_2(n))$ and $f_2 \in \mathcal{O}(n)$. Comparing the categories we have

$$n < n\log_2(n).$$

This means algorithm one is slower.

Example 8.22

Suppose algorithm one has time-complexity function $f_1(n) = 25\sqrt{n} + 10n\log_2(n)$ and algorithm two has time-complexity function $f_2(n) = 100n\log_2(n) + 20n + 50$. Which algorithm is faster?

First we have to find the big-$\mathcal{O}$ categories of each algorithm. It is easy to see that $f_1 \in \mathcal{O}(n\log_2(n))$ and $f_2 \in \mathcal{O}(n\log_2(n))$. Both algorithms are in the same big-$\mathcal{O}$ category. This means we consider these two algorithms to be equally fast.

8.6 PROBLEMS

Question 8.1 *Determine the dominant operation for all the algorithms presented in the problem section of chapter Introduction to Algorithms.*

Question 8.2 *Let $f(n) = 7n$ and $g(n) = n$. Show that f is $\mathcal{O}(g)$.*

Question 8.3 *Let $f(n) = 25n + 500$ and $g(n) = n$. Show that f is $\mathcal{O}(g)$.*

Question 8.4 *For $f(n) = 20\sqrt{n} + 40n^2 + 500n\log_2(n)$ find g such that f is $\mathcal{O}(g)$.*

Question 8.5 *For $f(n) = 800n^2 + 40n^3 + 50\log_2(n)$ find g such that f is $\mathcal{O}(g)$.*

Question 8.6 *Suppose that algorithm one has the time complexity function f_1 and algorithm two has time complexity function f_2. Given the time complexity functions below, which algorithm would you expect to be faster, or would you expect the algorithms to be equally fast.*
(a) $f_1(n) = n^2$ and $f_2(n) = n^3$,
(b) $f_1(n) = n^3 + n^2$ and $f_2(n) = 20n^2$,
(c) $f_1(n) = 500n^3$ and $f_2(n) = 10n^3$,
(d) $f_1(n) = 10n^2 - 20n$ and $f_2(n) = 20n^3 - 50n^2$,
(e) $f_1(n) = 15n^5$ and $f_2(n) = 500n^4 + 500n^3$,
(f) $f_1(n) = 10000n^2$ and $f_2(n) = 0.001n^3$,
(g) $f_1(n) = n^4 + 500n^2$ and $f_2(n) = n^5 + 500n^2$,
(h) $f_1(n) = 2n$ and $f_2(n) = 7000n$.

Question 8.7 *Suppose that algorithm one has the time complexity function f_1 and algorithm two has time complexity function f_2. Given the time complexity functions below, which algorithm would you expect to be faster, or would you expect the two algorithms to be equally fast.*
(a) $f_1(n) = 10\log_2(n) + 1000$ and $f_2(n) = 50\sqrt{n}$,
(b) $f_1(n) = 10000n\log_2(n)$ and $f_2(n) = 0.00001n$,
(c) $f_1(n) = 0.00001\sqrt{n}$ and $f_2(n) = 100000\log_2(n) + 50$,
(d) $f_1(n) = 500n^2$ and $f_2(n) = 2^n$,
(e) $f_1(n) = 500\sqrt{n} + \log_2(n)$ and $f_2(n) = 500\log_2(n) + \sqrt{n}$,
(f) $f_1(n) = 0.1n + 10\sqrt{n}$ and $f_2(n) = 0.1\sqrt{n} + 10\log_2(n)$,

Question 8.8 *Arrange the following algorithms in order of increasing growth rate:*

(a) 10^n (b) $\sqrt{n}$ (c) $n^{1.5}$

Question 8.9 *Arrange the following algorithms in order of increasing growth rate:*

(a) 2^n (b) $n^2 \log_2(n)$ (c) 2^{2^n}

Question 8.10 *Arrange the following algorithms in order of increasing growth rate:*

(a) $2^{n \log_2(n)}$ (b) 2^n (c) $2^{\log_2(n)}$

Question 8.11 *What is the dominant operation in the following algorithm? What is the time complexity function of this algorithm?*

1. Input $x_1, x_2, \ldots, x_n$.
2. $i \longleftarrow 1$
3. $order \longleftarrow true$
4. **While** $i < n$ and $order = \text{true}$ **do**
    4.1. **If** $x_i > x_{i+1}$ **then**
        4.1.1. $order \longleftarrow \text{false}$
    4.2. $i \longleftarrow i + 1$
5. **If** $order = \text{true}$ **then**
    5.1. Output "Numbers are in order."
    else
    5.2. Output "Numbers are out of order."

Question 8.12 *Given two $n \times n$ matrices A and B the following algorithm performs matrix operation. (An $n \times n$ matrix is an $n \times n$ array of numbers labeled a_{ij} where $1 \leq i \leq n$ and $1 \leq j \leq n$. Thus the matrix consists of n^2 numbers.) What is the dominant operation in the following algorithm? What is the time complexity function of this algorithm?*

1. Input a_{ij} and b_{ij} for $1 \leq i \leq n$ and $1 \leq j \leq n$.
2. **For** $i = 1$ to n **do**
    2.1. **For** $j = 1$ to n **do**
        2.1.1. $sum \longleftarrow 0$
        2.1.2. **For** $k = i$ to n **do**
            2.1.2.1 $sum \longleftarrow sum + x_{ik} \times x_{kj}$
        2.1.3 $c_{ij} \longleftarrow sum$
3. Output c_{ij}.

Question 8.13 *The following is Warshall's algorithm. What is the dominant operation in the following algorithm? What is the time complexity function of this algorithm?*

1. Input adjacency matrix (a_{ij}).
2. **For** $i = 1$ to n **do**
    2.1. $a_{ii} = 1$
3. **For** $i = 1$ to n **do**
    3.1 **For** $j = 1$ to n **do**
        3.1.1 **If** $a_{ij} = 1$ **then**
            3.1.1.1 **For** $k = 1$ to n **do**
                3.1.1.1.1 $a_{jk} \longleftarrow a_{jk} + a_{ik}$ (Boolean Addition)
4. Output (a_{ij}).

CHAPTER 9

Graph Theory

Graph theory is used to model many kinds of real world situations, relations, and processes that include communication networks, data organization, and computational devices. Many of the tasks that programmers, computer engineers, and computer scientists need to solve are made much easier by using graph theory. Developing algorithms to handled graphs is very important in computer science. This makes it necessary for you to understand the basic ideas of graph theory.

9.1 BASIC DEFINITIONS

The word graph in the phrase graph theory is very different from what the word graph means when we looked at functions. Now the word graph refers to something else entirely.

Definition 9.1 *A **graph** is a set of points called **vertices**[1] (or **nodes**) and a set of lines called **edges** such that each edge is attached to a vertex at each end.*

Very often the vertices are labeled to tell them apart. Two vertices that are connected by an edge are called **adjacent** vertices. An edge is said to be **incident** to the vertex to which it is attached. The various edges that are connected to a single vertex are said to be incident to each other. The number of vertices of a graph is called the **order** of the graph. A graph must have one vertex but it is allowable for a graph to have no edges. A graph that has no edges is sometimes called a **null graph**.

Example 9.1

A graph with six vertices and seven edges. This graph has order six. The vertices are labeled with capital letters.

A graph that is entirely connected is called a **connected** graph. However, it is possible

---

[1] The singular of vertices is vertex.

for a graph to be **disconnected** and have several pieces. These pieces are called **components** of the graph.

Example 9.2

A disconnected order nine graph with three components. One component consists of only a single vertex.

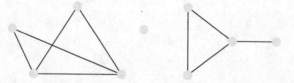

Graphs are made up of a set of vertices, called the **vertex set**, and a set of edges, called an **edge set**. Vertices and edges can be labeled many different ways, but for now we will label vertices with capital letters. In order to write down the edge set we have to have a way of labeling edges. In this book the edges will usually be labeled by the two vertices connected by the edge. For example, an edge that connects vertices A and B can be labeled by $\overline{AB}$ or $\overline{BA}$. Usually we will stick with alphabetical ordering and write this edge as $\overline{AB}$.

Example 9.3

Find the vertex set and the edge set of the following graph.

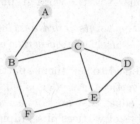

The vertex set is given by

$$\mathcal{V} = \{A, B, C, D, E, F\}.$$

Notice that for this graph vertex A is adjacent to vertex B since they are connected by an edge. Similarly vertex B is adjacent to vertices A, C, and F, and so on. The edge set is given by

$$\mathcal{E} = \{\overline{AB}, \overline{BC}, \overline{BF}, \overline{CD}, \overline{CE}, \overline{DE}, \overline{EF}\}.$$

In this graph the edges $\overline{AB}$, $\overline{BC}$, and $\overline{BF}$ are all incident to each other since they are all connected to vertex B. Similarly, the edges $\overline{CD}$ and $\overline{DE}$ are incident to each other since they are both connected to vertex D, and so on.

Often we use dictionary ordering[2] when writing down the edge set, as we did in the last example. If there is more than one edge that connects two different vertices these edges are called **parallel edges**. An edge that connects a vertex to itself is called a **loop**. A graph is called a **simple graph** if it does not have any parallel edges or loops.

---

[2] The technical term for this kind of ordering is lexicographical ordering.

Example 9.4

Does this graph have any parallel edges? Does it have any loops? Find the vertex and edge sets of the following graph.

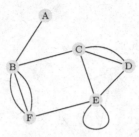

This graph has two sets of parallel edges. The first set of parallel edges are the three edges that connect the vertices B and F and the second set of parallel edges are the two edges that connect vertices C and D. There is also one loop, the edge that connects vertex E to itself. Notice that this edge indeed looks like a loop. The vertex set of the graph is given by

$$\mathcal{V} = \{A, B, C, D, E, F\}.$$

The edge set of the graph is given by

$$\mathcal{E} = \{\overline{AB}, \overline{BC}, \overline{BF}, \overline{BF}, \overline{BF}, \overline{CD}, \overline{CD}, \overline{CE}, \overline{DE}, \overline{EE}, \overline{EF}\}.$$

Think back to the chapter on set theory. There we said that the sets $\{1, 7, 3\}$ and $\{1, 3, 1, 7\}$ were considered as the same set even though the element 1 is repeated in the second set. We said that repeated elements did not matter. In the edge set of the above example it looks like the elements $\overline{BF}$ is repeated three times and $\overline{CD}$ is repeated twice. However, each of the elements $\overline{BF}$ represents a different edge and each of the elements $\overline{CD}$ represents a different edge. The edge set is a set that contains the edges, we just happened to use the same name for different edges. As long as you understand this hopefully there will be no confusion. Also consider the element $\overline{EE}$. This is the loop, it connects the vertex E to itself.

A simple graph is called a **complete graph** if every vertex of the graph is adjacent to (joined by an edge to) every other vertex of the graph.

Example 9.5

The complete graphs up to order five.

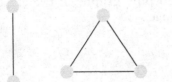

Sometimes, depending on what the graph represents, the graph's edges need to have a direction assigned to them. A graph whose edges have a direction assigned to them is called a **directed graph**. Very similar to the idea of directed graphs is the idea of **weighted**

graphs. A weighted graph is a graph where every edge has a number called a **weight** assigned to it. Usually the weights are positive numbers. These "weights" can represent things like distances or costs. We will have more to say about weighed graphs in chapter 10.

Example 9.6

An example of a directed graph.

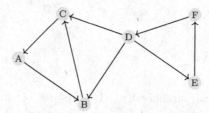

In this example the edge between vertices A and B goes only in one direction, from vertex A to vertex B. We can see this by noticing the little arrow on the edge $\overline{AB}$. The other edges are similar.

Example 9.7

An example of a weighted graph.

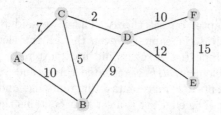

In this example the weight of edge $\overline{AB}$ is 10, the weight of $\overline{AC}$ is 7, and so on.

Of course, it should be obvious to you that there are literally an infinite number of different ways of drawing a graph. We can put the vertices anywhere we want. We can draw the edges any way we want. All that matters when drawing a graph is which vertices are adjacent to each other, that is, which vertices are connected to each other by an edge. The locations of the vertices and the lengths of the edges do not matter. Consider the following example. A little thought should convince you that even though these graphs are drawn differently that they all are really the same. Graphs that are drawn differently but still represent the same graph are called **isomorphic graphs**. If two graphs are isomorphic, then we consider them to be exactly the same.

Example 9.8

Four isomorphic graphs.

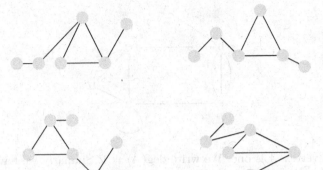

Example 9.9

Two isomorphic graphs.

Both graphs have edges $\overline{AD}$ and $\overline{BC}$. In the graph to the left it appears that these two edges cross each other. You may think that this means the two edges are somehow connected. They are not. The graph on the right is the correct way to imagine these two edges. However, it is easier to draw the graph on the left, so you will often see edges that are not connected draw crossing over each other. Just remember that this does not mean the edges are connected. But you should realize that some graphs cannot be drawn unless we cross edges that are not connected.

The **degree of a vertex** is defined to be the number of edges that are connected to it. Loops are counted twice in the degree of a vertex since one side of the edge is "out of" the vertex and the other side is "into" the vertex. A vertex that is part of a graph but that has no edges connected to it is called an **isolated** vertex. With the degree of the vertices defined this way we have the following theorem.

Theorem 9.1 *Given a graph $\mathcal{G}$, the sum of the degrees of the vertices of $\mathcal{G}$ is equal to twice the number of edges of $\mathcal{G}$. Using set notation we can write this as*

$$\sum_{v \in \mathcal{V}} \deg(v) = 2|\mathcal{E}|$$

where $\mathcal{V}$ is the vertex set and $\mathcal{E}$ is the edge set.

Example 9.10

Find the degrees of the vertices of the following graph.

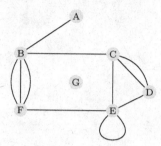

The degree of vertex A is one. We write $\deg(A) = 1$. Similarly we have $\deg(B) = 5$, $\deg(C) = 4$, $\deg(D) = 3$, and $\deg(F) = 4$. But what about vertex E? There are four edges connected to vertex E but the edge $\overline{EE}$ is connected to vertex E twice. In this case we count the edge $\overline{EE}$ twice to give us $\deg(E) = 5$. And what about vertex G? There are no edges connected to G so we have $\deg(G) = 0$. It is worth mentioning that even though in this picture vertex G is placed "inside" the graph it does not have to be. We could have placed vertex G anywhere.

Example 9.11

Suppose a graph has four vertices and eight edges. Two vertices have degree two and one vertex has degree three. What is the degree of the fourth vertex?

First we will label the vertices. Vertex v_1 and v_2 each have degree 2 and vertex v_3 has degree 3. We want to find what degree vertex v_4 has. Also, since there are eight edges we have $|\mathcal{E}| = 8$. Using the above theorem we have

$$\sum_{v \in \mathcal{V}} \deg(v) = 2 \, |\mathcal{E}|$$

$$\Rightarrow \deg(v_1) + \deg(v_2) + \deg(v_3) + \deg(v_4) = 2 \, |\mathcal{E}|$$

$$\Rightarrow 2 + 2 + 3 + \deg(v_4) = 2(8)$$

$$\Rightarrow 7 + \deg(v_4) = 16$$

$$\Rightarrow \deg(v_4) = 9.$$

Thus vertex four must have degree nine.

A **path** is a sequence of vertices such that each vertex is adjacent to the next vertex. Thus a path consists of a sequence of edges. By going along the path you travel across this sequence of edges. The number of edges traveled across is called the **path length**. For us a path may include repeated vertices and edges. Some books may define this a little differently and say a path cannot include repeated edges. If we want to talk about a path that has no repeated edges we will say that explicitly. A path that starts and ends at the same vertex is called a **circuit** (graph theory). There are different ways to write a path. Below are some of the acceptable ways to write down a path.

$$ABFBCD$$
$$A - B - F - B - C - D$$
$$A \rightarrow B \rightarrow F \rightarrow B \rightarrow C \rightarrow D$$
$$A, B, F, B, C, D$$

Example 9.12

Some examples of paths.

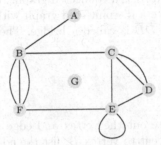

The sequence $ABFEC$ is a path of length four. You start at vertex A and then move along the edge that connects vertex A to vertex B. Once at vertex B you move along one of the edges connecting vertex B to vertex F and arrive at vertex F. Which edge you take is not specified so it does not matter. Then you move from vertex F to vertex E along the edge that connects them, and then finally you move to vertex C along the edge that connects those vertices. Some other examples of paths are $FBCDEFBA$ or $EEDCDEC$. Some examples of circuits are $BFEDCB$ or $FBFECBF$.

Example 9.13

Paths on a directed graph.

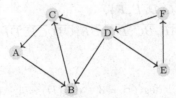

A path on a directed graph must follow the arrows, so the path $CABC$ is acceptable but $CDBC$ is not since to go from C to D would require going along the edge $\overline{CD}$ in the wrong direction.

An edge that keeps a graph connected is called a **bridge**.

Example 9.14

An example of a bridge.

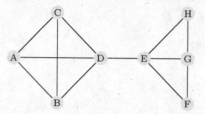 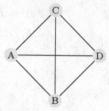 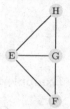

Consider the graph on the left. It is a connected graph. However, if we were to remove edge $\overline{DE}$ then we would have a disconnected graph with two components, as shown on the right. Thus the edge $\overline{DE}$ is called a bridge. The graph $\mathcal{G}$ with the edge $\overline{DE}$ deleted is written as $\mathcal{G} - \overline{DE}$.

Example 9.15

For the following graph write out the vertex and edge sets, find the degree of each vertex, list the vertices adjacent to vertex B, list the edges incident to vertex D, list the loops (if any), list the parallel edges (if any), and list the bridges (if any).

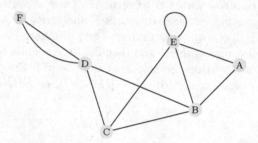

The vertex and edge sets are given by

$$\mathcal{V} = \{A, B, C, D, E, F\},$$
$$\mathcal{E} = \{\overline{AB}, \overline{AE}, \overline{BC}, \overline{BD}, \overline{BE}, \overline{CD}, \overline{CE}, \overline{DF}, \overline{DF}, \overline{EE}\}.$$

The degrees of the vertices are

$$\deg(A) = 2, \quad \deg(B) = 4, \quad \deg(C) = 3, \quad \deg(D) = 4, \quad \deg(E) = 5, \quad \deg(F) = 2.$$

The vertices adjacent to vertex B are A, C, D, and E. The edges incident to vertex D are $\overline{BD}$, $\overline{CD}$, $\overline{DF}$, and $\overline{DF}$. There is one loop, $\overline{EE}$, and there are two parallel edges $\overline{DF}$ and $\overline{DF}$. There are no bridges.

9.2 EULERIAN AND SEMI-EULERIAN GRAPHS

If $\mathcal{G}$ is a connected graph then an **Euler path** on graph $\mathcal{G}$ is a path that includes every edge of $\mathcal{G}$ exactly once. Vertices may be repeated on an Euler path. An **Euler circuit** is an Euler path that starts and ends at the same vertex. A connected graph that has an Euler circuit is called **Eulerian graph**. A connected graph that has an Euler path but no Euler

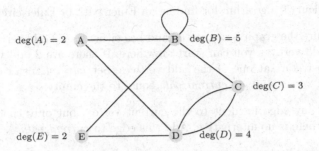

Figure 9.1 A semi-Eulerian graph. The path $BBADCDEBC$ is an example of an Euler path. Notice, two vertices have odd degree.

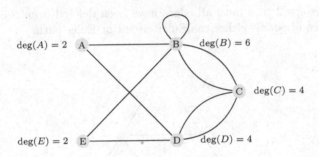

Figure 9.2 An Eulerian graph. The path $CDCBBADEBC$ is an example of an Euler circuit. Notice, all vertices have even degree.

circuit is called a **semi-Eulerian graph**. A graph that is not connected cannot have either an Euler path or an Euler circuit.

The graph in Fig. 9.1 is an example of a semi-Eulerian graph. One example of an Euler path is $BBADCDEBC$. Another is $CDCBBADEB$. A semi-Eulerian graph will generally have many different Euler paths. The graph in Fig. 9.2 is an example of a Eulerian graph. An example of an Euler circuit is $CDCBBADEBC$. Another is $ABBCBEDCDA$. An Eulerian graph will generally have many different Euler circuits.

Now notice that the graph that has an Euler path has exactly two vertices with an odd degree and the rest of the vertices have an even degree. In the graph that has an Euler circuit all vertices have an even degree. There is a theorem about this.

Theorem 9.2 *Let $\mathcal{G}$ be a connected graph, then*

1. *if every vertex in $\mathcal{G}$ has an even degree then $\mathcal{G}$ has an Euler circuit,*

2. *if $\mathcal{G}$ has exactly two vertices of odd degree (with the rest being of even degree), then $\mathcal{G}$ has an Euler path, and*

3. *if $\mathcal{G}$ has more than two vertices of odd degree, then $\mathcal{G}$ does not have either an Euler circuit or an Euler path.*

This theorem is not hard to prove, but we will not prove it here. Instead we will give an algorithm for finding the Euler circuit or path called **Fleury's algorithm**. The algorithm that we will give works, but it is not what we would use to write a computer program, so we will not write it in pseudocode. The pseudocode version is given later.

Algorithm: Fleury's algorithm for finding an Euler path or Euler circuit.

1. Make sure the graph is connected and has either 0 or 2 odd vertices. If there are 0 odd vertices you can start anywhere. If there are 2 odd vertices then you must start at one of the odd vertices. Set *current_vertex* equal to the starting vertex and set *current_path* equal to the empty set.

2. Choose any edge incident to the current vertex, but only choose a bridge only if there is no alternative. Add this edge to *current_path*.

3. Set *current_vertex* equal to the vertex at the other end of the edge. (If the edge is a loop the *current_vertex* does not change.)

4. Delete the edge and then delete any isolated vertices from the graph.

5. Repeat steps 2 to 4 until all edges have been deleted from the graph. The final *current_path* is either the Euler circuit or Euler path.

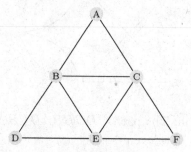

Figure 9.3 A graph $\mathcal{G}$ with all even degree vertices.

Let use Fleury's algorithm on the graph $\mathcal{G}$ given in Fig. 9.3. This graph $\mathcal{G}$ has the vertex set $\mathcal{V}$ and edge set $\mathcal{E}$,

$$V = \{A, B, C, D, E, F\},$$
$$\mathcal{E} = \{\overline{AB}, \overline{AC}, \overline{BC}, \overline{BD}, \overline{BE}, \overline{CE}, \overline{CF}, \overline{DE}, \overline{EF}\}.$$

Notice, every vertex here is even so we choose to simply start at vertex A, the first vertex in our vertex set $\mathcal{V}$.

(a) Starting at vertex A we see both edges $\overline{AB}$ or edge $\overline{AC}$ are incident to A. We will simply chose whichever edge is listed first in the edge set $\mathcal{E}$. We follow $\overline{AB}$ to vertex B which becomes our new *current_vertex* and then delete edge $\overline{AB}$ from $\mathcal{E}$ and add it to *current_path*. See Fig. 9.4(a). We are left with

$$V = \{A, B, C, D, E, F\},$$
$$\mathcal{E} = \{\overline{AC}, \overline{BC}, \overline{BD}, \overline{BE}, \overline{CE}, \overline{CF}, \overline{DE}, \overline{EF}\},$$
$$current\_vertex = B,$$
$$current\_path = \{\overline{AB}\}.$$

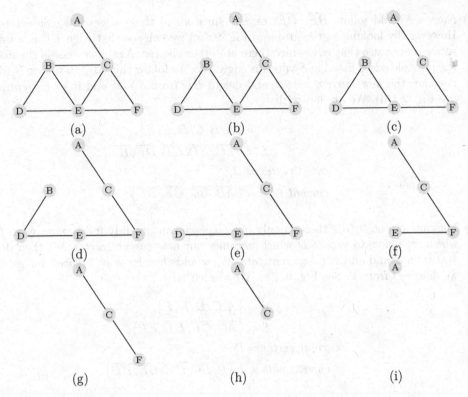

Figure 9.4 The steps of Fleury's algorithm for the graph given in Fig. 9.3

(b) Next we could follow $\overline{BC}$, $\overline{BD}$, or $\overline{BE}$ since all of these edges are incident to B and none is a bridge. We simply choose the edge listed next in the edge set $\mathcal{E}$ which is edge $\overline{BC}$. We follow $\overline{BC}$ to vertex C which becomes the new *current_vertex*, and delete $\overline{BC}$ from $\mathcal{E}$ and add it to *current_path*. See Fig. 9.4(b). We are left with

$$\mathcal{V} = \{A, B, C, D, E, F\},$$
$$\mathcal{E} = \{\overline{AC}, \overline{BD}, \overline{BE}, \overline{CE}, \overline{CF}, \overline{DE}, \overline{EF}\},$$
$$current\_vertex = C,$$
$$current\_path = \{\overline{AB}, \overline{BC}\}.$$

(c) Next we could follow $\overline{AC}$, $\overline{CE}$, or $\overline{CF}$ since all of these edges are incident to C. However, by looking at the graph in Fig. 9.4(b) we can see that edge $\overline{AC}$ is a bridge so we cannot use that edge since there are other choices. Again we choose the first of the possible edges listed in $\mathcal{E}$ which is edge $\overline{CE}$. We follow this edge to vertex E which becomes the new *current_vertex*, and delete $\overline{CE}$ from $\mathcal{E}$ and add it to *current_path*. See Fig. 9.4(c). We are left with

$$\mathcal{V} = \{A, B, C, D, E, F\},$$
$$\mathcal{E} = \{\overline{AC}, \overline{BD}, \overline{BE}, \overline{CF}, \overline{DE}, \overline{EF}\},$$
$$current\_vertex = E,$$
$$current\_path = \{\overline{AB}, \overline{BC}, \overline{CE}\}.$$

(d) Next we could follow $\overline{BE}$, $\overline{DE}$, or $\overline{EF}$ since all of these edges are incident to E. However, by looking at the graph in Fig. 9.4(c) we can see that edge $\overline{EF}$ is a bridge so we cannot use that edge since there are other choices. Again we choose the first of the possible edges listed in $\mathcal{E}$ which is edge $\overline{BE}$. We follow this edge to vertex B which becomes the new *current_vertex*, and delete $\overline{BE}$ from $\mathcal{E}$ and add it to *current_path*. See Fig. 9.4(d). We are left with

$$\mathcal{V} = \{A, B, C, D, E, F\},$$
$$\mathcal{E} = \{\overline{AC}, \overline{BD}, \overline{CF}, \overline{DE}, \overline{EF}\},$$
$$current\_vertex = B,$$
$$current\_path = \{\overline{AB}, \overline{BC}, \overline{CE}, \overline{BE}\}.$$

(e) Looking at Fig. 9.4(d) there is only one edge we can possibly follow now, edge $\overline{BD}$, which we follow to vertex D which becomes our new *current_vertex*. We then delete $\overline{BD}$ from $\mathcal{E}$ and add it to *current_path*. We would then have an isolated vertex B so we delete B from $\mathcal{V}$. See Fig. 9.4(e). We are left with

$$\mathcal{V} = \{A, C, D, E, F\},$$
$$\mathcal{E} = \{\overline{AC}, \overline{CF}, \overline{DE}, \overline{EF}\},$$
$$current\_vertex = D,$$
$$current\_path = \{\overline{AB}, \overline{BC}, \overline{CE}, \overline{BE}, \overline{BD}\}.$$

(f) Looking at Fig. 9.4(e) there is only one edge we can possibly follow now, edge $\overline{DE}$, which we follow to vertex E which becomes our new *current_vertex*. We then delete $\overline{DE}$ from $\mathcal{E}$ and add it to *current_path*. We would then have an isolated vertex D so we delete D from $\mathcal{V}$. See Fig. 9.4(f). We are left with

$$\mathcal{V} = \{A, C, E, F\},$$
$$\mathcal{E} = \{\overline{AC}, \overline{CF}, \overline{EF}\},$$
$$current\_vertex = E,$$
$$current\_path = \{\overline{AB}, \overline{BC}, \overline{CE}, \overline{BE}, \overline{BD}, \overline{DE}\}.$$

(g) Looking at Fig. 9.4(f) there is only one edge we can possibly follow now, edge $\overline{EF}$, which we follow to vertex F which becomes our new *current_vertex*. We then delete $\overline{EF}$ from $\mathcal{E}$ and add it to *current_path*. We would then have an isolated vertex E so we delete E from $\mathcal{V}$. See Fig. 9.4(g). We are left with

$$\mathcal{V} = \{A, C, F\},$$
$$\mathcal{E} = \{\overline{AC}, \overline{CF}\},$$
$$current\_vertex = F,$$
$$current\_path = \{\overline{AB}, \overline{BC}, \overline{CE}, \overline{BE}, \overline{BD}, \overline{DE}, \overline{EF}\}.$$

(h) Looking at Fig. 9.4(g) there is only one edge we can possibly follow now, edge $\overline{CF}$, which we follow to vertex C which becomes our new *current_vertex*. We then delete $\overline{CF}$ from $\mathcal{E}$ and add it to *current_path*. We would then have an isolated vertex F so we delete F from $\mathcal{V}$. See Fig. 9.4(h). We are left with

$$\mathcal{V} = \{A, C\},$$
$$\mathcal{E} = \{\overline{AC}\},$$
$$current\_vertex = C,$$
$$current\_path = \{\overline{AB}, \overline{BC}, \overline{CE}, \overline{BE}, \overline{BD}, \overline{DE}, \overline{EF}, \overline{CF}\}.$$

(i) Looking at Fig. 9.4(h) there is only one edge we can possibly follow now, edge $\overline{AC}$, which we follow to vertex A which becomes our new *current_vertex*. We then delete $\overline{AC}$ from $\mathcal{E}$ and add it to *current_path*. We would then have an isolated vertex C so we delete C from $\mathcal{V}$. See Fig. 9.4(i). We are left with

$$\mathcal{V} = \{A\},$$
$$\mathcal{E} = \{\ \},$$
$$current\_vertex = A,$$
$$current\_path = \{\overline{AB}, \overline{BC}, \overline{CE}, \overline{BE}, \overline{BD}, \overline{DE}, \overline{EF}, \overline{CF}, \overline{AC}\}.$$

At this point all edges have been deleted from the graph and we are now done. The *current_path* set is given by

$$current\_path = \{\overline{AB}, \overline{BC}, \overline{CE}, \overline{BE}, \overline{BD}, \overline{DE}, \overline{EF}, \overline{CF}, \overline{AC}\}$$

which we use to construct our path. The path we have is given by

$$A - B - C - E - B - D - E - F - C - A$$

which is clearly an Euler circuit.

Why did we say that this algorithm is not what we would use to write a computer program? Yes, we do have to make choices in the algorithm; which vertex to start at and which path to choose. But in both of those cases we simply chose the first appropriate choice in the set $\mathcal{V}$ or $\mathcal{E}$. The real problem is that in step two we had to determine if an edge was a bridge or not. As humans we can look at the graph and see quickly if an edge is or is not a bridge. But a computer cannot "see" the graph the same way humans can and so cannot easily determine if an edge is a bridge or not. Fleury's algorithm was written in such a way that certain choices in it need human level intelligence. However, we can change Fleury's algorithm a little bit so that a computer can implement it.

The below algorithm is a modification of Fleury's algorithm that is more appropriate for a computer. We will now trace this algorithm for the same graph we used when explaining Fleury's algorithm. This will allow you to compare the two algorithms. Hopefully you will start to see why this algorithm is better for implementing in a computer. It is modified in such a way that there is no need to decide if a given edge is a bridge or not.

In order to make it easier to trace this algorithm we will relabel the edges in the graph $\mathcal{G}$ from Fig. 9.3 in line with the notation used in the algorithm,

$$\overline{AB} = e_1, \ \overline{AC} = e_2, \ \overline{BC} = e_3, \ \overline{BD} = e_4, \ \overline{BE} = e_5, \ \overline{CE} = e_6, \ \overline{CF} = e_7, \ \overline{DE} = e_8, \ \overline{EF} = e_9.$$

However, we will maintain the same vertex labels as before. See Fig. 9.5. Also, instead of doing a complete trace we will consider the four steps in 3.3 as a singe step. We will also leave off the final output column so the table is not too wide to fit on the page.

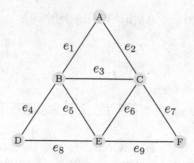

Figure 9.5 The graph from Fig. 9.3 with the edges relabeled.

Algorithm: Modified Fleury's Algorithm to find an Euler circuit for an Eulerian graph.

1. Input Eulerian graph $\mathcal{G}$ with vertex set $\mathcal{V} = \{v_1, v_2, \ldots, v_m\}$ and edges set $\mathcal{E} = \{e_1, e_2, \ldots, e_n\}$.
2. *current_path* $\longleftarrow v_1$; *unused_edges* $\longleftarrow \mathcal{E}$
3. **While** *unused_edges* $\neq \varnothing$ **do**
   3.1. *insertion_point* $\longleftarrow$ first vertex in *currpent_path* with unused edges incident to it
   3.2. $v \longleftarrow$ *insertion_point*; *new_path* $\longleftarrow v$
   3.3. **Repeat**
      3.3.1. $e \longleftarrow$ first element of *unused_edges* incident to v
      3.3.2. $v \longleftarrow$ vertex adjacent to v via edge e
      3.3.3. *new_path* $\longleftarrow$ *new_path*, e, v
      3.3.4. *unused_edges* $\longleftarrow$ *unused_edges* $- \{e\}$
      until no element of *unused_edges* is incident to v
   3.4. *current_path* $\longleftarrow$ (*current_path* before *insertion_point*), *new_path*, (*current_path* after *insertion_point*)
4. Output *current_path*.

The trace of the modified Fleury's algorithm for the graph in Fig. 9.5 is shown in Table 9.1. At the end of the trace we have *unused_edges* $= \varnothing$ and so the **while-do** loop ends and we move to step four, which outputs the *current_path*,

$$Ae_1Be_4De_8Ee_6Ce_7Fe_9Ee_5Be_3Ce_2A.$$

Writing *current_path* as we have written graph paths in this chapter we have

$$A - B - D - E - C - F - E - B - C - A.$$

This is indeed an Euler circuit for the graph $\mathcal{G}$ though it is different from the one we had obtained using Fleury's algorithm. Notice how this algorithm was constructed so we never needed to decide if a loop was a bridge or not. Instead, we kept finding new paths using *new_path* and inserting them into *current_path*.

This algorithm needs to be changed just a little bit if you have a semi-Eulerian graph and want to find an Euler path instead. Recall, a semi-Eulerian graph has two vertices with odd degree with the rest of the vertices having even degree.

Step	current_path	insertion_point	e	v	new_path	unused_edges
2	A	-	-	-	-	$\{e_1, e_2, \ldots, e_9\}$
3.1	A	A	-	-	-	$\{e_1, e_2, \ldots, e_9\}$
3.2	A	A	-	A	A	$\{e_1, e_2, \ldots, e_9\}$
3.3.1–4	A	A	e_1	B	Ae_1B	$\{e_2, e_3, \ldots, e_9\}$
3.3.1–4	A	A	e_3	C	Ae_1Be_3C	$\{e_2, e_4, e_5, e_6, e_7, e_8, e_9\}$
3.3.1–4	A	A	e_2	A	$Ae_1Be_3Ce_2A$	$\{e_4, e_5, e_6, e_7, e_8, e_9\}$
3.4	$Ae_1Be_3Ce_2A$	A	e_2	A	$Ae_1Be_3Ce_2A$	$\{e_4, e_5, e_6, e_7, e_8, e_9\}$
3.1	$Ae_1Be_3Ce_2A$	B	e_2	A	$Ae_1Be_3Ce_2A$	$\{e_4, e_5, e_6, e_7, e_8, e_9\}$
3.2	$Ae_1Be_3Ce_2A$	B	e_2	B	B	$\{e_4, e_5, e_6, e_7, e_8, e_9\}$
3.3.1–4	$Ae_1Be_3Ce_2A$	B	e_4	D	Be_4D	$\{e_5, e_6, e_7, e_8, e_9\}$
3.3.1–4	$Ae_1Be_3Ce_2A$	B	e_8	E	Be_4De_8E	$\{e_5, e_6, e_7, e_9\}$
3.3.1–4	$Ae_1Be_3Ce_2A$	B	e_5	B	$Be_4De_8Ee_5B$	$\{e_6, e_7, e_9\}$
3.4	$Ae_1Be_4De_8Ee_5Be_3Ce_2A$	B	e_5	B	$Be_4De_8Ee_5B$	$\{e_6, e_7, e_9\}$
3.1	$Ae_1Be_4De_8Ee_5Be_3Ce_2A$	E	e_5	B	$Be_4De_8Ee_5B$	$\{e_6, e_7, e_9\}$
3.2	$Ae_1Be_4De_8Ee_5Be_3Ce_2A$	E	e_5	E	E	$\{e_6, e_7, e_9\}$
3.3.1–4	$Ae_1Be_4De_8Ee_5Be_3Ce_2A$	E	e_6	C	Ee_6C	$\{e_7, e_9\}$
3.3.1–4	$Ae_1Be_4De_8Ee_5Be_3Ce_2A$	E	e_7	F	Ee_6Ce_7F	$\{e_9\}$
3.3.1–4	$Ae_1Be_4De_8Ee_5Be_3Ce_2A$	E	e_9	E	$Ee_6Ce_7Fe_9E$	$\{\}$
3.4	$Ae_1Be_4De_8Ee_6Ce_7Fe_9$ $Ee_5Be_3Ce_2A$	E	e_9	E	$Ee_6Ce_7Fe_9E$	$\{\}$

Table 9.1 Trace of modified Fleury's algorithm for graph in Fig. 9.5.

Algorithm: Finds the Euler path for a semi-Eulerian graph.

1. Let v_1 and v_2 be the two vertices with odd degree. Insert a new edge, called e_1, between v_1 and v_2. Our new graph $\mathcal{G}$ is now an Eulerian graph.

2. Let the vertex set of $\mathcal{G}$ be $\mathcal{V} = \{v_1, v_2, \ldots, v_m\}$ and let the edge set be $\mathcal{E} = \{e_1, e_2, \ldots, e_n\}$.

3. Apply the algorithm for finding the Euler circuit of an Eulerian graph to $\mathcal{G}$.

4. Remove v_1, e_1 from the beginning of the Euler circuit you obtain. What remains is an Euler path joining the two vertices of odd degree.

9.3 MATRIX REPRESENTATIONS OF GRAPHS

The way we have drawn graphs up till now is a wonderful way for us to visualize and understand graphs. But of course a computer cannot visualize or understand graphs the same way humans do. We need to be able to represent a graph in a way that a computer can handle. One way of doing this is simply to use the vertex set and edge set to represent the graph. This way of representing a graph is good enough for the modified Fleury's algorithm. Another common way to represent graphs is to use matrices.

First we explain what a matrix is. A **matrix** is a rectangular array of numbers, generally written enclosed in square brackets. For example,

$$\begin{bmatrix} 7 & 4 & -3 \\ 2 & -9 & 5 \end{bmatrix}, \quad \begin{bmatrix} 2 & -6 \\ 9 & 4 \\ 11 & 1 \\ 6 & 8 \end{bmatrix}, \quad \text{and} \quad \begin{bmatrix} 2 & -4 & 6 & 7 \\ -1 & 8 & 0 & 10 \\ 5 & 2 & -3 & 7 \\ -2 & -4 & 1 & 3 \end{bmatrix}$$

are all examples of matrices. The size of the matrix is indicated with the number of rows and columns, usually written as (rows)×(columns). For example, the sizes of the above matrices are 2×3 (read "two by three"), 4×2 (read "four by two"), and 4×4 (read "four by four"). A matrix that has the same number of rows and columns is called a **square matrix**. The entries in the matrix are sometimes called matrix elements and are denoted with variables that have two subscripts, the first subscript for the row number and the second subscript for the column number. Here a 3×3 matrix is written using variables with subscripts,

$$\begin{bmatrix} a_{11} & a_{12} & a_{13} \\ a_{21} & a_{22} & a_{23} \\ a_{31} & a_{32} & a_{33} \end{bmatrix}.$$

It is important to remember that the row subscript comes first and the column subscript comes second, a_{rc}. Thus, for the 2×3 matrix

$$\begin{bmatrix} 7 & 4 & -3 \\ 2 & -9 & 5 \end{bmatrix}.$$

we would have

$$a_{11} = 7, \quad a_{12} = 4, \quad a_{13} = -3,$$
$$a_{21} = 2, \quad a_{22} = -9, \quad a_{23} = 5.$$

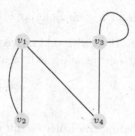

Figure 9.6 An undirected graph used to construct an adjacency matrix.

Since computers can handle arrays of numbers easily, a graph is represented in computers by using a special matrix. The most common kind of matrix is called an **adjacency matrix**. Suppose we have a graph whose vertex set is given by $\mathcal{V} = \{v_1, v_2, \ldots, v_n\}$. The elements of the adjacency matrix are defined by

$$a_{ij} = \text{the number of edges from vertex } v_i \text{ to vertex } v_j.$$

For the moment we will label vertices with v_i instead of the usual capital letters to make it easier for you to see how the adjacency matrix is made. Consider the graph in Fig. 9.6. Since there are no loops going from v_1 to v_1 we have $a_{11} = 0$. Since there are two edges from v_1 to v_2 we have $a_{12} = 2$. Since there is one edge from v_1 to v_3 we have $a_{13} = 1$. Since there is one edge from v_1 to v_4 we have $a_{14} = 1$. Thus we have just found the first row in the below table. The other rows are similar.

	v_1	v_2	v_3	v_4
v_1	0	2	1	1
v_2	2	0	0	0
v_3	1	0	1	1
v_4	1	0	1	0

This table would give us the adjacency matrix

$$\begin{bmatrix} 0 & 2 & 1 & 1 \\ 2 & 0 & 0 & 0 \\ 1 & 0 & 1 & 1 \\ 1 & 0 & 1 & 0 \end{bmatrix}.$$

But notice something, for an undirected graph the number of edges from vertex v_i to vertex v_j is exactly the same as the number of edges from vertex v_j to vertex v_i. This is because they are exactly the same edges. Therefore, in the adjacency matrix we have $a_{ij} = a_{ji}$. Matrices that satisfy this property are called **symmetric matrices**. They are said to be symmetric "across the diagonal." The diagonal of the matrix is represented by the matrix elements a_{11}, a_{22}, a_{33}, and so on. Notice that the lower left triangular region is a reflection of the upper right triangular region. It is easy to see this matrix is symmetrical,

$$\begin{bmatrix} 0 & 2 & 1 & 1 \\ 2 & 0 & 0 & 0 \\ 1 & 0 & 1 & 1 \\ 1 & 0 & 1 & 0 \end{bmatrix}.$$

Because the matrix is symmetrical sometimes you will see the adjacency matrix written down with only the diagonal and one of the triangular regions filled out. The **upper triangular adjacency matrix** and **lower triangular adjacency matrix** are given by

$$\begin{bmatrix} 0 & 2 & 1 & 1 \\ & 0 & 0 & 0 \\ & & 1 & 1 \\ & & & 0 \end{bmatrix} \quad \text{and} \quad \begin{bmatrix} 0 & & & \\ 2 & 0 & & \\ 1 & 0 & 1 & \\ 1 & 0 & 1 & 0 \end{bmatrix}$$

respectively. Since both of these matrices give all the information needed you should not be surprised if you sometimes see an adjacency matrix written like either of these matrices.

Example 9.16

Draw the graph for the following adjacency matrix

$$\begin{bmatrix} 0 & 1 & 0 & 1 & 1 \\ 1 & 0 & 1 & 1 & 1 \\ 0 & 1 & 0 & 1 & 1 \\ 1 & 1 & 1 & 0 & 1 \\ 1 & 1 & 1 & 1 & 0 \end{bmatrix}.$$

Here is one possible answer:

Example 9.17

Find the adjacency matrix for the following graph.

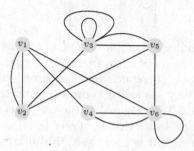

We begin by filling out a table for this graph. Since this is an undirected graph we can simply fill out the diagonal and one of the triangular regions.

	v_1	v_2	v_3	v_4	v_5	v_6
v_1	0	2	0	1	0	1
v_2		0	1	0	1	0
v_3			2	0	2	0
v_4				0	0	2
v_5					0	1
v_6						1

It is then easy to write down the adjacency matrix using this table. In fact, this table already gives the upper triangular adjacency matrix.

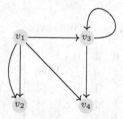

Figure 9.7 A directed graph used to construct an adjacency matrix.

Finally, we will take a quick look at the adjacency matrix for a directed graph. We will use the same graph as before in Fig. 9.6, but we will now add directions to each edge, see Fig. 9.7. This will allow you to compare between the adjacency matrices for undirected and directed graphs. We follow basically the same procedure as before, except that now we have to specify which vertices the edges are "from" and which vertices the edges go "to." In the below table we will say the edges go from the vertices along the left and to the vertices along the top. So, the terms in the adjacency matrix of a directed graph are given by

$$a_{ij} = \text{the number of edges from vertex } v_i \text{ to vertex } v_j.$$

Since the first index in the subscript of a_{ij} represents the row then the v_i are the "from" vertices and since the second index in the subscript of a_{ij} represents the column then the v_j are the "to" vertices. The adjacency matrix for a directed graph will not usually be symmetric.

For example, we have two edges going from v_1 to v_2 making $a_{12} = 2$. We have one edge going from v_1 to v_3 so $a_{13} = 1$. We have one edge going from v_1 to v_4 making $a_{14} = 1$. The loop goes from v_3 and then back to v_3 giving us $a_{33} = 1$. And finally, we have one edge going from v_3 to v_4 giving us $a_{34} = 1$. The rest of the terms are zero. This gives us the following table

		to			
		v_1	v_2	v_3	v_4
	v_1	0	2	1	1
	v_2	0	0	0	0
from	v_3	0	0	1	1
	v_4	0	0	0	0

which in turn gives us the adjacency matrix

$$\begin{bmatrix} 0 & 2 & 1 & 1 \\ 0 & 0 & 0 & 0 \\ 0 & 0 & 1 & 1 \\ 0 & 0 & 0 & 0 \end{bmatrix}.$$

Notice that this adjacency matrix is no longer symmetric.

Also, notice that there is no reason the "from" vertices need to be along the left and the "to" vertices need to be along the top. These can be switched easily. Sometime the adjacency matrix is defined differently than we have defined it here. Whenever you read about or see an adjacency matrix used you should look carefully at how it is defined.

Finally, we will make one point. While there was nothing wrong with what we did if our graph modeled some sort of real situation most of the time we only care that we can go from vertex v_1 to vertex v_2, we do not actually care that there are two (or more) different paths to from v_1 to v_2. Thus directed graphs usually do not have parallel edges that go in the same direction. An adjacency matrix for a directed graph where parallel edges are not allowed would only have zeros and ones in it.

9.4   REACHABILITY FOR DIRECTED GRAPHS

Given a directed graph it is often important to know if you can reach some vertex when you are starting at some other vertex. In this case we do not care if there are multiple paths from one vertex to another so we will say that parallel edges in the same direction are not allowed. Recall that in the last section we defined the terms in the adjacency matrix by

$$a_{ij} = \text{the number of edges from vertex } v_i \text{ to vertex } v_j.$$

If a directed graphs does not have parallel edges this becomes

$$a_{ij} = 1 \text{ if there is an edge from } v_i \text{ to } v_j,$$
$$a_{ij} = 0 \text{ if there is no edge from } v_i \text{ to } v_j.$$

Take a look at Fig. 9.8. This is a directed graph with no parallel edges going in the same direction. Suppose you are sitting at vertex v_1 and you want to know if you can get to v_4.

Figure 9.8 A directed graph.

By looking at the directed graph we can see that there is a path from v_1 to v_4, namely the path

$$v_1 \rightarrow v_2 \rightarrow v_3 \rightarrow v_4$$

Now suppose we wanted to go from vertex v_2 to v_1. By looking at the graph we can see that there is no way to actually get to v_1. When the directed graphs are small it is easy for us to look at the graph and answer this question. It is not so easy for computers. Nor is it easy for humans if the directed graph is large.

If there exists a path from v_i to v_j we say that v_j is **reachable** from v_i. We also say that v_i is reachable from v_i. This just means that if you are standing at a vertex you can reach that vertex by not moving. You can think of this as a path of length zero. A matrix that tells us when we can reach v_j from v_i is called a **reachability matrix**. In a reachability matrix

$$a_{ij} = 1 \text{ if there is a path from } v_i \text{ to } v_j,$$
$$a_{ij} = 0 \text{ if there is no path from } v_i \text{ to } v_j.$$

Notice how this looks almost the same as the definition for a_{ij} in the adjacency matrix. The only difference is that instead of saying *edge* it says *path*. We can think of the adjacency matrix as a "one step" reachability matrix. In other words, an adjacency matrix tells us what is reachable by a path of length one. A reachability matrix tells us what is reachable by a path of any length, including length zero.

Example 9.18

Using the definition, find the reachability matrix for the directed graph in Fig. 9.8.

Just by looking we can see that if we are on vertex v_1 we can reach v_1 (just by staying where we are), v_2, v_3, and v_4. This gives $a_{11} = 1$, $a_{12} = 1$, $a_{13} = 1$, and $a_{14} = 1$. If we are on vertex v_2 we can reach v_2 (again by staying where we are), v_3 and v_4 but can not reach v_1. This gives $a_{21} = 0$, $a_{22} = 1$, $a_{23} = 1$, and $a_{24} = 1$. Similarly, if we are on vertex v_3 we can reach v_3, v_4, and v_2 but not v_1. This gives $a_{31} = 0$, $a_{32} = 1$, $a_{33} = 1$, and $a_{34} = 1$. And finally, if we are on vertex v_4 we can reach v_4, v_2, and v_3 but not v_1. This gives $a_{41} = 0$, $a_{42} = 1$, $a_{42} = 1$, and $a_{44} = 1$. Putting this together gives us the following reachability matrix,

$$\begin{bmatrix} 1 & 1 & 1 & 1 \\ 0 & 1 & 1 & 1 \\ 0 & 1 & 1 & 1 \\ 0 & 1 & 1 & 1 \end{bmatrix}.$$

If we have the adjacency matrix for a directed graph there is an algorithm, called Warshall's algorithm, that produces the reachability matrix. In order to use Warshall's algorithm we first have to recall Boolean addition. Boolean addition is defined on the set $\{0, 1\}$ by

$$0 + 0 = 0,$$
$$0 + 1 = 1,$$
$$1 + 0 = 1,$$
$$1 + 1 = 1.$$

Similar to the case of Fleury's algorithm, we will give a version of Warshall's algorithm that is easier to understand before giving a more formal pseudocode version.

> **Algorithm: Warshall's algorithm to determine if there exists a directed path from v_i to v_j.**
>
> Given the adjacency matrix (a_{ij}) of a directed graph:
>
> 1. Change all entries along the diagonal to 1.
>
> 2. Using the matrix obtained in step one, for every row where in column one there is a 1, add row one of this matrix to the row that has the value 1 using Boolean addition.
>
> 3. Using the matrix obtained in step two, for every row where in column two there is a 1, add row two of this matrix to the row that has the value 1 using Boolean addition.
>
> 4. Continue for all columns.

Using Warshall's algorithm we will find the reachability matrix for the directed graph in Fig. 9.8. We begin by writing down the adjacency matrix for this graph,

$$\begin{bmatrix} 0 & 1 & 0 & 0 \\ 0 & 0 & 1 & 0 \\ 0 & 0 & 0 & 1 \\ 0 & 1 & 0 & 0 \end{bmatrix}.$$

(a) In step 1 we change all the entries along the diagonal to one, which gives us the matrix

$$\begin{bmatrix} 1 & 1 & 0 & 0 \\ 0 & 1 & 1 & 0 \\ 0 & 0 & 1 & 1 \\ 0 & 1 & 0 & 1 \end{bmatrix}.$$

We do this because every vertex is reachable from itself. In other words, for each i we can reach v_i from v_i by simply staying where we are so we make $a_{ii} = 1$.

(b) In step 2 we look at column one of the matrix we obtained above. Column one is written in gray,

$$\begin{bmatrix} 1 & 1 & 0 & 0 \\ 0 & 1 & 1 & 0 \\ 0 & 0 & 1 & 1 \\ 0 & 1 & 0 & 1 \end{bmatrix}.$$

For every row where in column one there is a 1, we add row one of this matrix to the row that has the value 1 using Boolean addition. Row one of this matrix is given by $[1, 1, 0, 0]$. Notice we put commas in between the elements of the row just to make it clear.

Since row one is the only row that has a 1 in it in column one then we add row one to itself using Boolean addition. This means we add each term in the row to itself. Because we are using Boolean addition this does not change the row at all,

$$[1, 1, 0, 0] + [1, 1, 0, 0] = [1 + 1, 1 + 1, 0 + 0, 0 + 0]$$
$$= [1, 1, 0, 0].$$

(c) In step 3 we look at column two of the matrix we obtained above. Column two is written in gray,

$$\begin{bmatrix} 1 & 1 & 0 & 0 \\ 0 & 1 & 1 & 0 \\ 0 & 0 & 1 & 1 \\ 0 & 1 & 0 & 1 \end{bmatrix}.$$

For every row where in column two there is a 1, we add row two of this matrix to the row that has the value 1 using Boolean addition. Row two of this matrix is given by $[0, 1, 1, 0]$ and there are three rows in this column that have a value 1, row one, row two, and row four.

First we will add row two and row one,

$$[0, 1, 1, 0] + [1, 1, 0, 0] = [0 + 1, 1 + 1, 1 + 0, 0 + 0]$$
$$= [1, 1, 1, 0]$$

which becomes our new row one. The Boolean addition of row two with row two gives us

$$[0, 1, 1, 0] + [0, 1, 1, 0] = [0 + 0, 1 + 1, 1 + 1, 0 + 0]$$
$$= [0, 1, 1, 0]$$

which is exactly row two again. Last we will add row two and row four,

$$[0, 1, 1, 0] + [0, 1, 0, 1] = [0 + 0, 1 + 1, 1 + 0, 0 + 1]$$
$$= [0, 1, 1, 1]$$

which becomes our new row four. Putting this altogether we have obtained the matrix

$$\begin{bmatrix} 1 & 1 & 1 & 0 \\ 0 & 1 & 1 & 0 \\ 0 & 0 & 1 & 1 \\ 0 & 1 & 1 & 1 \end{bmatrix}.$$

(d) Next we look at column three of the matrix we obtained above. Column three is written in gray,

$$\begin{bmatrix} 1 & 1 & 1 & 0 \\ 0 & 1 & 1 & 0 \\ 0 & 0 & 1 & 1 \\ 0 & 1 & 1 & 1 \end{bmatrix}.$$

For every row where in column three there is a 1, we add row three of this matrix to the row that has the value 1 using Boolean addition. Row three of this matrix is given by $[0, 0, 1, 1]$. Since every row in this column has a value 1, we add row three to every row using Boolean addition. We will not give the details, only the results

$$\text{row three} + \text{row one} = [1, 1, 1, 1],$$
$$\text{row three} + \text{row two} = [0, 1, 1, 1],$$
$$\text{row three} + \text{row three} = [0, 0, 1, 1],$$
$$\text{row three} + \text{row four} = [0, 1, 1, 1]$$

giving us matrix

$$\begin{bmatrix} 1 & 1 & 1 & 1 \\ 0 & 1 & 1 & 1 \\ 0 & 0 & 1 & 1 \\ 0 & 1 & 1 & 1 \end{bmatrix}.$$

(e) Next we look at column four of the matrix we obtained above. Column four is written in gray,

$$\begin{bmatrix} 1 & 1 & 1 & 1 \\ 0 & 1 & 1 & 1 \\ 0 & 0 & 1 & 1 \\ 0 & 1 & 1 & 1 \end{bmatrix}.$$

For every row where in column four there is a 1, we add row four of this matrix to the row that has the value 1 using Boolean addition. Row four of this matrix is given by $[0, 1, 1, 1]$. Since every row in this column has a value 1, we add row four to every row using Boolean addition. We will not give the details, only the results

$$\text{row four} + \text{row one} = [1, 1, 1, 1],$$
$$\text{row four} + \text{row two} = [0, 1, 1, 1],$$
$$\text{row four} + \text{row three} = [0, 1, 1, 1],$$
$$\text{row four} + \text{row four} = [0, 1, 1, 1]$$

giving us matrix

$$\begin{bmatrix} 1 & 1 & 1 & 1 \\ 0 & 1 & 1 & 1 \\ 0 & 1 & 1 & 1 \\ 0 & 1 & 1 & 1 \end{bmatrix}.$$

(f) Since we have now finished all four columns we are done. The Reachability matrix is given by the matrix we obtained above,

$$\begin{bmatrix} 1 & 1 & 1 & 1 \\ 0 & 1 & 1 & 1 \\ 0 & 1 & 1 & 1 \\ 0 & 1 & 1 & 1 \end{bmatrix}.$$

Since $a_{21} = 0$ there is no path from v_2 to v_1, since $a_{31} = 0$ there is no path from v_3 to v_1, and since $a_{41} = 0$ there is no path from v_4 to v_1. However, there are paths in all the other cases. Notice how this is exactly what we would expect from looking at Fig. 9.8.

This procedure is far easier to understand than looking directly at the pseudocode for Warshall's algorithm, which is given below. However, when writing a program for a computer the pseudocode below is what you would base your program on.

Algorithm: Warshall's algorithm to determine if there exists a directed path from v_i to v_j. (Pseudocode version.)

  1. Input adjacency matrix (a_{ij}).
  2. **For** $i = 1$ to n **do**
     2.1. $a_{ii} = 1$
  3. **For** $i = 1$ to n **do**
     3.1 **For** $j = 1$ to n **do**
        3.1.1 **If** $a_{ij} = 1$ **then**
           3.1.1.1 **For** $k = 1$ to n **do**
              3.1.1.1.1 $a_{jk} \longleftarrow a_{jk} + a_{ik}$ (Boolean Addition)
  4. Output (a_{ij}).

9.5 PROBLEMS

Question 9.1 *For the graphs in Fig. 9.9 do the following:*

(a) write out the vertex and edge set,
(b) find the order of the graph,
(c) find the degree for each vertex,
(d) list the vertices that are adjacent to vertex B,
(e) list the edges incident to vertex C,
(f) list the loops (if any),
(g) list the parallel edges (if any), and
(h) list the bridges (if any).

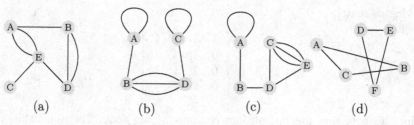

(a)   (b)   (c)   (d)

Figure 9.9 The graphs for questions 9.1, 9.11, and 9.18.

Question 9.2 *Assume you have a graph $\mathcal{G}$ with the following vertex and edge set,*

$$\mathcal{V} = \{A, B, C, D, E\}$$
$$\mathcal{E} = \{\overline{AA}, \overline{AD}, \overline{AE}, \overline{BC}, \overline{BD}, \overline{CD}, \overline{CE}, \overline{DD}\}.$$

Which of the following pairs of vertices are adjacent?

(a) A and B	(d) B and C	(g) E and A
(b) A and C	(e) B and D	(h) E and C
(c) A and D	(f) B and E	(i) E and D

Question 9.3 *Assume you have a graph $\mathcal{G}$ with the following vertex and edge set,*

$$\mathcal{V} = \{A, B, C, D, E, F\}$$
$$\mathcal{E} = \{\overline{AB}, \overline{AC}, \overline{BC}, \overline{BC}, \overline{DE}, \overline{DF}, \overline{DF}, \overline{DD}\}.$$

Is $\mathcal{G}$ connected? If not, how many components does $\mathcal{G}$ have?

Question 9.4 *Assume you have a graph $\mathcal{G}$ with the following vertex and edge set,*

$$\mathcal{V} = \{A, B, C, D, E\}$$
$$\mathcal{E} = \{\overline{AA}, \overline{AD}, \overline{AE}, \overline{BC}, \overline{BD}, \overline{CD}, \overline{CE}, \overline{DD}\}.$$

Find the degree of all the vertices of $\mathcal{G}$. Find all loops, all parallel edges, and all bridges.

Question 9.5 *Assume you have a graph $\mathcal{G}$ with the following vertex and edge set,*

$$\mathcal{V} = \{A, B, C, D, E, F\}$$
$$\mathcal{E} = \{\overline{AA}, \overline{AD}, \overline{AE}, \overline{BC}, \overline{BD}, \overline{BD}, \overline{CD}, \overline{EF}, \overline{FF}\}.$$

Find the degree of all the vertices of $\mathcal{G}$. Find all loops, all parallel edges, and all bridges.

Question 9.6 *Assume you have a graph $\mathcal{G}$ with the following vertex and edge set,*

$$\mathcal{V} = \{A, B, C, D, E, F\}$$
$$\mathcal{E} = \{\overline{AD}, \overline{AE}, \overline{BC}, \overline{BD}, \overline{BD}, \overline{CD}, \overline{EF}, \overline{FF}\}.$$

If the edges $\overline{AD}$ and $\overline{EF}$ were removed, how many components would the resulting graph have?

Question 9.7 *Suppose you have a graph with five vertices. Two vertices have degree 3, two vertices have degree 5, and one vertex has degree 2. How many edges does the graph have?*

Question 9.8 *Suppose you have a graph with eight vertices, four vertices have degree 4, two vertices have degree 3, one vertex has degree 6, and one vertex has degree 8. How many edges does the graph have?*

Question 9.9 *Suppose a graph has four vertices and five edges. One vertex has degree 1, one vertex has degree 2, and one vertex has degree 3. What is the degree of the remaining vertex?*

Question 9.10 *Suppose a graph has six vertices and twelve edges. There are two vertices of degree 4, one vertex with degree 2, one vertex with degree 3, and one vertex with degree 6. What is the degree of the remaining vertex?*

Question 9.11 *Determine if the graphs in Fig. 9.9 are Eulerian, semi-Eulerian, or neither.*

Question 9.12 *Determine if the graphs in Fig. 9.10 are Eulerian, semi-Eulerian, or neither.*

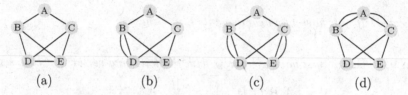

    (a)   (b)   (c)   (d)

Figure 9.10   The graphs for questions 9.12 and 9.19.

Question 9.13 *Assume you have a graph $\mathcal{G}$ with the following vertex and edge set,*

$$\mathcal{V} = \{A, B, C, D, E, F\}$$
$$\mathcal{E} = \{\overline{AA}, \overline{AD}, \overline{AE}, \overline{BC}, \overline{BD}, \overline{BD}, \overline{CD}, \overline{EF}, \overline{FF}\}.$$

Is $\mathcal{G}$ Eulerian, semi-Eulerian, or neither?

Question 9.14 *Use Fleury's algorithm to find an Euler path on the below graph.*

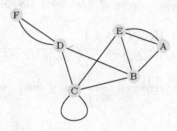

Question 9.15 *Assume you have a graph $\mathcal{G}$ with the following vertex and edge set,*

$$\mathcal{V} = \{A, B, C, D, E, F\}$$
$$\mathcal{E} = \{\overline{AA}, \overline{AD}, \overline{AE}, \overline{BC}, \overline{BD}, \overline{BD}, \overline{CD}, \overline{EF}, \overline{FF}\}.$$

Use Fleury's algorithm to find an Euler path on $\mathcal{G}$.

Question 9.16 *Assume you have a graph $\mathcal{G}$ with the following vertex and edge set,*

$$\mathcal{V} = \{A, B, C, D\}$$
$$\mathcal{E} = \{\overline{AB}, \overline{AC}, \overline{BC}, \overline{BC}, \overline{BD}, \overline{CD}\}.$$

Show that $\mathcal{G}$ is an Eulerian graph. Use the modified Fleury's algorithm for finding the Euler circuit on an Eulerian graph starting at vertex A. It helps to relabel the edges as $\overline{AB} = e_1, \overline{AC} = e_2, \overline{BC} = e_3$, and so on.

Question 9.17 *Assume you have a graph $\mathcal{G}$ with the following vertex and edge set,*

$$\mathcal{V} = \{A, B, C, D, E\}$$
$$\mathcal{E} = \{\overline{AB}, \overline{AC}, \overline{BC}, \overline{BD}, \overline{BE}, \overline{CD}, \overline{CE}, \overline{DE}, \overline{DE}\}.$$

Show that $\mathcal{G}$ is an Eulerian graph. Use the modified Fleury's algorithm for finding the Euler circuit on an Eulerian graph starting at vertex A. It helps to relabel the edges as $\overline{AB} = e_1, \overline{AC} = e_2, \overline{BC} = e_3$, and so on.

Question 9.18 *Find the adjacency matrices for the graphs in Fig. 9.9.*

Question 9.19 *Find the adjacency matrices for the graphs in Fig. 9.10.*

Question 9.20 *Assume you have a graph $\mathcal{G}$ with the following vertex and edge set,*

$$\mathcal{V} = \{A, B, C, D, E\}$$
$$\mathcal{E} = \{\overline{AA}, \overline{AD}, \overline{AE}, \overline{BC}, \overline{BD}, \overline{CD}, \overline{CE}, \overline{DD}\}.$$

Find the adjacency matrix for $\mathcal{G}$.

Question 9.21 *Assume you have a graph $\mathcal{G}$ with the following vertex and edge set,*

$$\mathcal{V} = \{A, B, C, D, E, F\}$$
$$\mathcal{E} = \{\overline{AB}, \overline{AC}, \overline{BC}, \overline{BC}, \overline{DD}, \overline{DE}, \overline{DF}, \overline{DF}, \}.$$

Find the adjacency matrix for $\mathcal{G}$.

Question 9.22 *Assume you have a graph $\mathcal{G}$ with the following vertex and edge set,*

$$\mathcal{V} = \{A, B, C, D, E, F\}$$
$$\mathcal{E} = \{\overline{AD}, \overline{AE}, \overline{BC}, \overline{BD}, \overline{BD}, \overline{CD}, \overline{EF}, \overline{FF}\}.$$

Find the adjacency matrix for $\mathcal{G}$.

Question 9.23 *Find the time complexity for Warshall's algorithm.*

Question 9.24 *Use Warshall's algorithm to find the reachability matrix for the below graph.*

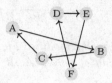

Question 9.25 *Use Warshall's algorithm to find the reachability matrix for the below graph.*

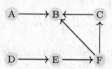

CHAPTER 10

Trees

Trees are a special kind of graph that show up everywhere in computer science. Operating systems use a tree structure for directories, folders, and files. Tree structures are used to store data that has a hierarchical structure. These data structures are easy to search and sort using standard algorithms. Trees are also used to process the syntax of computer languages. Recall, we used expression trees in logic and Boolean algebra to understand the order of operations, write truth tables, and draw circuit diagrams. Also, in many applications it is important to find a tree that comes from a directed or weighted graph. A good understanding of tree basics is essential in computer science.

10.1   BASIC DEFINITIONS

We will now study a particular type of graph, called a tree, that shows up often in computer science and computer engineering. First, recall that a path is a sequence of vertices where each vertex is adjacent to the next vertex and a circuit is a path that starts and ends at the same vertex. Though other books may have slightly different definitions, for us repeated edges or vertices are allowed in paths and circuits unless we say they are not allowed.

A **cycle** is a circuit that includes at least one edge; has no repeated edges; and also has no repeated vertices except for the first and last vertices. Thus a cycle is a circuit where repeated edges and vertices, except for the first and last vertex, are not allowed. The length of a cycle is the number of edges in the cycle. If a cycle has length n is called an n-cycle. A 1-cycle must be a loop and a 2-cycle must be a pair of parallel edges.

Example 10.1

Examples of paths, circuits, and cycles.

Both $ABCE$ and $ABCDCE$ are examples of paths and both $BCEFB$ and $BCDCB$ are examples of circuits. This graph has three cycles: $BCEFB$, $BCDEFB$, and $CDEC$. Of course it is possible to write these cycles in different ways. For example, the cycle $BCEFB$ could also be written as $CEFBC$ or as $FBCEF$ or as $ECBFE$, and so on.

A **tree** is a connected graph with no cycles. Trees are called trees because, well, they look like trees. Keeping with the tree analogy, a vertex of a tree that has degree one is often called a **leaf**. Trees have several properties that should be obvious just by looking at examples of trees. If T is a tree then the following properties are true;

- if T has n vertices then it has $n - 1$ edges,
- given any two vertices in T there is only one path between the vertices that does not repeat any vertices or edges,
- inserting an edge between any two vertices in T produces a new graph that has a cycle,
- every edge of T is a bridge. That means that if any edge is removed it produces a disconnected graph.

We will prove only the first of these properties below, but by looking at some examples of trees the other properties should be clear.

Example 10.2

Four examples of trees.

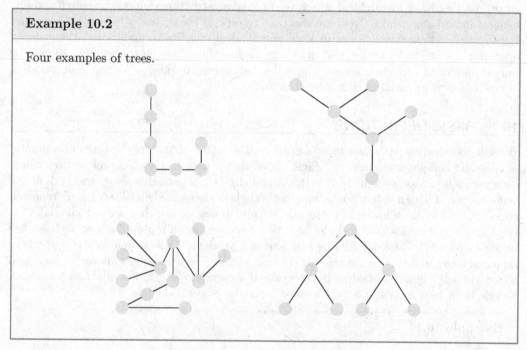

Take a careful look at the trees shown in the last example. Count the number of vertices and edges in each tree. You should notice that there is always one more vertex than there are edges. This is something that is always true for trees. The proof of this theorem is not very difficult so we will give it. It is also a very nice example of **induction**. Induction is an technique often used for proving certain kinds of statements and is widely used in computer science. That is why we give the proof of this theorem.

Theorem 10.1 *Any tree with n vertices has $n - 1$ edges. (Another way of saying this is that any tree with n edges has $n + 1$ vertices.)*

Proof: Suppose T_1 is a tree with one vertex. Since a tree cannot have any loops we know that there are no edges in tree T_1. Thus T_1 has 1 vertex and 0 edges.

Next suppose we have a tree T_2 with two vertices v_1 and v_2. Since trees are connected we know that there is an edge e connecting the two vertices. Remove v_2 and e. We are left with a tree that has 1 vertex, and by the last paragraph we know this tree has 0 edges. Now add back v_2 and e to get back $\mathcal{T}_2$, which increases the number of vertices by one and the number of edges by one. We now know that $\mathcal{T}_2$ has 2 vertices and 1 edge.

Next suppose we have a tree $\mathcal{T}_3$ with three vertices v_1, v_2, and v_3. Since trees are connected we know that there is an edge e connecting v_3 and some other vertex. Remove v_3 and e. We are left with a tree that has 2 vertices, and by the last paragraph we know it must have 2 vertices and 1 edge. Now add back v_3 and e to get back $\mathcal{T}_3$, which increases the number of vertices by one and the number of edges by one. We now know that $\mathcal{T}_3$ has 3 vertices and 2 edges.

And of course if we were to continue for a tree $\mathcal{T}_4$ the argument would be exactly the same. As it would be for $\mathcal{T}_5$, $\mathcal{T}_6$, and so on. But instead of continuing forever we assume that trees $\mathcal{T}_n$ with n vertices have $n-1$ edges. We then do a step called an *inductive step* where we use this to prove that a tree $\mathcal{T}_{n+1}$ has $n+1$ vertices and n edges. We now state the inductive step.

Suppose we have a tree $\mathcal{T}_{n+1}$ with vertices $v_1, \ldots, v_{n+1}$. Since trees are connected we know that there is an edge e connecting v_{n+1} with some other vertex. Remove v_{n+1} and e. We are left with a tree that has n vertices. By the assumption we know that this tree has n vertices and $n-1$ edges. Now add back v_{n+1} and e to get back $\mathcal{T}_{n+1}$, which clearly has $n+1$ vertices and n edges.

Since this inductive step works for any n we have proved the theorem. □

A **rooted tree** is a tree where one special vertex is called the **root**. Rooted trees are usually drawn with the root at the top and the rest of the tree "hanging" from the root. The vertex directly above a given vertex is called the **parent** of the given vertex. Similarly, the vertex directly below a given vertex is called the **child** of the given vertex. Recalling that leaves are vertices with degree one, leaves on rooted trees turn out to be vertices with no children. A **strict binary tree** is a rooted tree such that every vertex which is not a leaf has exactly two children. A **binary tree** is a rooted tree such that every vertex which is not a leaf has at most two children. This means it is possible to have only one child.

Example 10.3

A rooted tree.

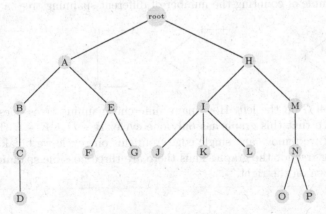

The vertex labeled "root" is the parent vertex of vertices A and H. Vertex A is the parent vertex of both B and E, and so on. Vertex D is the child of vertex C, and both vertices F and G are children of vertex E, and so on. Vertices D, F, G, J, K, L, O, and P are all leaves.

A **spanning tree** of a graph G is a tree with the same vertex set as G and with an edge set that is a subset of the edge set of G.

Example 10.4

Example of a graph G along with nine of its possible spanning trees.

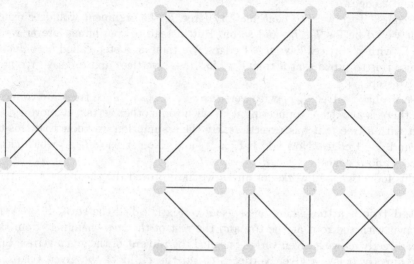

A graph G (left) along with nine possible spanning trees (right). Notice that each spanning tree has the same four vertices as G and has three edges from G. And, of course, each spanning tree is a tree.

In some cases it is not difficult to find out how many different spanning trees a graph has.

Example 10.5

A simple example of counting the number of different spanning trees a graph has.

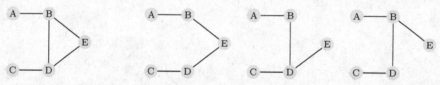

Consider graph G on the left. How many different spanning trees does it have? The key is to notice that this graph has only one cycle, $B - D - E - B$ that consists of three edges. If we remove any single edge from this one cycle we break the cycle and get a spanning tree for the graph. Thus there are three possible spanning trees, all of which are shown on the right.

Example 10.6

An example of counting the number of spanning trees a graph has.

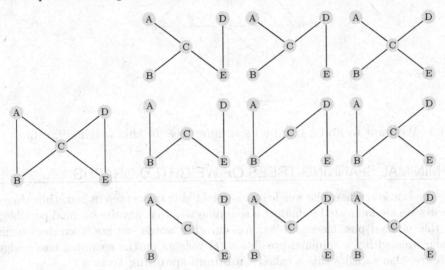

How many different spanning trees does the graph on the left have? This graph has two disjoint cycles, $A - B - C - A$ and $D - E - C - D$, each with three edges. If we remove exactly one edge from the first cycle and one edge from the second cycle we break each cycle and get a spanning three. There are three possible edges that we can remove from the first cycle and three possible edges that we can remove from the second cycle giving us a total of $3 \times 3 = 9$ possible ways to obtain a spanning tree. These nine different spanning trees are shown on the right. Notice, this is exactly an application of the multiplication principle from chapter 7.

Example 10.7

Counting the number of spanning trees a graph with two cycles which are not disjoint has.

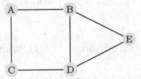

In the graph there are two cycles, $A - C - D - B - A$ and $B - E - D - B$. The cycle $A - C - D - B - A$ contains four edges and the cycle $B - E - D - B$ contains three edges. We may assume that, like before, the number of possible spanning trees is given by $4 \times 3 = 12$ but this is not quite right. Remember, before we were allowed to remove one edge from each cycle. However, now the two cycles share edge $\overline{BD}$. The option where we remove edge $\overline{BD}$ from the first cycle and edge $\overline{BD}$ from the second cycle does not give a tree and so is not allowed. So the number of possible spanning trees is given by $4 \times 3 - 1 = 12 - 1 = 11$. Try to draw all 11 spanning trees for this graph.

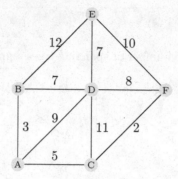

Figure 10.1 We want to find a minimal spanning tree for this weighted graph.

10.2   MINIMAL SPANNING TREES OF WEIGHTED GRAPHS

In many applications where the weights in a weighted graph represent something like costs or distances we are interested in finding a spanning tree that has the minimal possible total cost or the minimal possible total distance. In other words, we are interested in finding a spanning tree with the minimal possible total weights on the spanning tree's edges. A spanning tree that satisfies this is called a **minimal spanning tree**.

Let us consider the graph in Fig. 10.1. We want to find the minimal spanning tree for this graph. To do this we will use **Prim's algorithm**. We begin by listing the vertex set and edges set for $\mathcal{G}$,

$$\mathcal{V} = \{A, B, C, D, E, F\},$$
$$\mathcal{E} = \{\overline{AB}_3, \overline{AC}_5, \overline{AD}_9, \overline{BD}_7, \overline{BE}_{12}, \overline{CD}_{11}, \overline{CF}_2, \overline{DE}_7, \overline{DF}_8, \overline{EF}_{10}\}$$

where we have written the weight of each edge as a subscript to help us easily remember the weights.

Prim's algorithm is an example of a **greedy algorithm**. In a greedy algorithm the "least expensive" option is chosen at each step. Notice how in each step of the algorithm the link with the minimal weight is chosen.

Algorithm: Prim's algorithm to find a minimal cost spanning tree of a weighted graph.

1. Input vertex set $\mathcal{V} = \{v_1, v_2, \ldots, v_n\}$ and edge set $\mathcal{E} = \{e_1, e_2, \ldots, e_m\}$.
2. $\mathcal{T} \longleftarrow v_1$; $unused\_edges \longleftarrow \mathcal{E}$
3. **For** $i = 1$ to $n - 1$ **do**
   3.1 $e \longleftarrow$ first edge with minimal weight which is incident to exactly one vertex in $\mathcal{T}$
   3.2 $v \longleftarrow$ vertex incident to e which is not in $\mathcal{T}$
   3.3 $\mathcal{T} \longleftarrow \mathcal{T} \cup \{e, v\}$
   3.4 $unused\_edges \longleftarrow unused\_edges - \{e\}$
4. Output $\mathcal{T}$.

Instead of simply doing a trace of Prim's algorithm we will discuss each step to make sure you fully understand the algorithm. We can start with any vertex so we will simply start with the first vertex in $\mathcal{V}$ which is A. In step two we assign the first vertex to $\mathcal{T}$ which gives us $\mathcal{T} = \{A\}$ and we assign $\mathcal{E}$ to $unused\_edges$ so we have $unused\_edges = \mathcal{E}$. We then begin step three, which is a **for-do** loop which will loop five times, one fewer time than the number of vertices in the graph.

(a) We look for the first edge with minimal weight which is incident to exactly one vertex in $\mathcal{T}$. Since $\mathcal{T} = \{A\}$ we simply have to look among the edges incident to vertex A, which are $\overline{AB}_3$, $\overline{AC}_5$, and $\overline{AD}_9$. Clearly $\overline{AB}_3$ is the edge with the minimal weight. The vertex incident to $\overline{AB}_3$ which is not in $\mathcal{T}$ is clearly B. Thus for steps 3.1–4 we have the following assignments,

$$e \longleftarrow \overline{AB}_3,$$
$$v \longleftarrow B,$$
$$\mathcal{T} \longleftarrow \{A\} \cup \{\overline{AB}_3, B\},$$
$$unused\_edges \longleftarrow \mathcal{E} - \{\overline{AB}_3\}.$$

(b) Next we look for the first edge with minimum weight which is incident to exactly one vertex in $\mathcal{T}$. These vertices are A and B so we need to look among $\overline{AC}_5$, $\overline{AD}_9$, $\overline{BD}_7$, and $\overline{BE}_{12}$. Among these edges $\overline{AC}_5$ has the minimal weight. So for steps 3.1–4 we are left with the assignments

$$e \longleftarrow \overline{AC}_5,$$
$$v \longleftarrow C,$$
$$\mathcal{T} \longleftarrow \{A, \overline{AB}_3, B\} \cup \{\overline{AC}_5, C\},$$
$$unused\_edges \longleftarrow \{\overline{AD}_9, \overline{BD}_7, \overline{BE}_{12}, \overline{CD}_{11}, \overline{CF}_2, \overline{DE}_7, \overline{DF}_8, \overline{EF}_{10}\}.$$

(c) Next we again look for the first edge with minimum weight which is incident to exactly one vertex in $\mathcal{T}$. These vertices are A, B, and C so we now look among $\overline{AD}_9$, $\overline{BD}_7$, $\overline{BE}_{12}$, $\overline{CD}_{11}$, and $\overline{CF}_2$. Among these edges the one with the minimal weight is $\overline{CF}_2$. So for steps 3.1–4 we have the assignments

$$e \longleftarrow \overline{CF}_2,$$
$$v \longleftarrow F,$$
$$\mathcal{T} \longleftarrow \{A, \overline{AB}_3, B, \overline{AC}_5, C\} \cup \{\overline{CF}_2, F\},$$
$$unused\_edges \longleftarrow \{\overline{AD}_9, \overline{BD}_7, \overline{BE}_{12}, \overline{CD}_{11}, \overline{DE}_7, \overline{DF}_8, \overline{EF}_{10}\}.$$

(d) Next we again look for the first edge with minimum weight which is incident to exactly one vertex in $\mathcal{T}$. These vertices are A, B, C, and F so we now look among $\overline{AD}_9$, $\overline{BD}_7$, $\overline{BE}_{12}$, $\overline{CD}_{11}$, $\overline{DF}_8$, and $\overline{EF}_{10}$. The edge with the minimal wight is $\overline{BD}_7$. So for steps 3.1–4 we have the assignments

$$e \longleftarrow \overline{BD}_7,$$
$$v \longleftarrow D,$$
$$\mathcal{T} \longleftarrow \{A, \overline{AB}_3, B, \overline{AC}_5, C, \overline{CF}_2, F\} \cup \{\overline{BD}_7, D\},$$
$$unused\_edges \longleftarrow \{\overline{AD}_9, \overline{BE}_{12}, \overline{CD}_{11}, \overline{DE}_7, \overline{DF}_8, \overline{EF}_{10}\}.$$

(e) Next we again look for the first edge with minimum weight which is incident to exactly one vertex in $\mathcal{T}$. These vertices are A, B, C, F, and D so we now look among $\overline{BE}_{12}$, $\overline{DE}_7$, and $\overline{EF}_{10}$. The edge with the minimal weight is $\overline{DE}_7$. So for steps 3.1–4 we have the assignments

$$e \longleftarrow \overline{DE}_7,$$
$$v \longleftarrow E,$$
$$\mathcal{T} \longleftarrow \{A, \overline{AB}_3, B, \overline{AC}_5, C, \overline{CF}_2, F, \overline{BD}_7, D\} \cup \{\overline{DE}_7, E\},$$
$$unused\_edges \longleftarrow \{\overline{AD}_9, \overline{BE}_{12}, \overline{CD}_{11}, \overline{DF}_8, \overline{EF}_{10}\}.$$

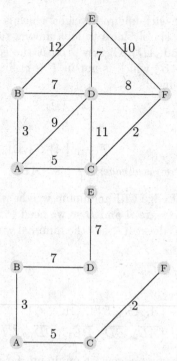

Figure 10.2 The weighted graph (top) with its minimal spanning tree (bottom) found using Prim's algorithm.

At this point we have gone through the **for-do** loop five times and we now move to step 4 which outputs the tree

$$\mathcal{T} = \{A, \overline{AB}_3, B, \overline{AC}_5, C, \overline{CF}_2, F, \overline{BD}_7, D, \overline{DE}_7, E\}.$$

See Fig. 10.2.

10.3   MINIMAL DISTANCE PATHS

Minimum distance problems are a little different from minimum spanning tree problems. Like the minimum spanning tree problem, we start with a weighted graph. In this case we usually interpret the weights as being distances. Given some vertex, say v_1, from a weighted graph we are interested in finding the minimum distance path from v_1 to any other vertex in the graph. Again, we will end up with a tree that connects v_1 with every other vertex. Notice, nothing is said about the overall sum of the edge weights that are included in the tree being a minimum. It is only the sum of the weights on the path from v_1 to each v_i must be the minimum of all the paths from v_1 to v_i. A spanning try that satisfies this is called called a **minimal distance spanning tree**. This can be done using Dijkstra's algorithm. Note, we may end up with a different tree than the one found using Prim's algorithm.

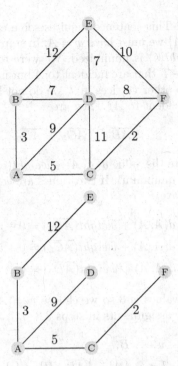

Figure 10.3 The weighted graph (top) with the spanning tree (bottom) found using Dijkstra's algorithm to solve the minimum distance problem from vertex A. Notice how different this spanning tree is from the one found using Prim's algorithm shown in Fig. 10.2.

Algorithm: Dijkstra's algorithm to find a minimal distance path.

1. Input graph $\mathcal{G}$ with vertex set $\mathcal{V} = \{v_1, v_2, \ldots, v_m\}$ and edges set $\mathcal{E} = \{e_1, e_2, \ldots, e_n\}$.
2. $\mathcal{T} \longleftarrow \{v_1\}$, $d(v_1, v_1) \longleftarrow 0$
3. **For** $j = 1$ to $n - 1$ **do**
   3.1. $w \longleftarrow$ first vertex $w \in \mathcal{T}$ for which $d(v_1, w) + weight(e)$ is minimized, where e ranges over edges in $\mathcal{G} - \mathcal{T}$ that are incident to w but not incident to another vertex in $\mathcal{T}$.
   3.2. $e \longleftarrow$ the first edge incident to w for which $d(v_1, w) + weight(e)$ is minimized.
   3.3. $v \longleftarrow$ the vertex in $\mathcal{G} - \mathcal{T}$ to which e is incident
   3.4. $\mathcal{T} \longleftarrow \mathcal{T} \cup \{e, v\}$
   3.5. $d(v_1, v) \longleftarrow d(v_1, w) + weight(e)$
4. Output $\mathcal{T}$.

We will use the same graph as before, the one shown in Fig. 10.1. We will again start with vertex A. And as before we will explain each step fully instead of doing a trace of the algorithm. Since we are starting at the vertex $v_1 = A$, in step two we set $\mathcal{T} = \{A\}$ and $d(A, A) = 0$. We then begin step 3 which is a **for-do** loop that executes a total of five times, one fewer than the number of vertices in our graph.

(a) In step 3.1 we are looking for the first vertex $w \in \mathcal{T}$ for which $d(A, w) + weight(e)$ is minimized, where e ranges over edges in $\mathcal{G} - \mathcal{T}$ that are incident to w but not incident to another vertex in $\mathcal{T}$. This sentence requires some explanation. First, it says that $w \in \mathcal{T}$ but since $\mathcal{T} = \{A\}$ we must have $w = A$. In step 3.2 we must also find the edge e where $d(A, A) + weight(e)$ is minimized. We were told the edges e we can choose from are the edges in $\mathcal{G} - \mathcal{T}$ that are incident to A but not incident to any other vertex in $\mathcal{T}$. Since there are no other vertices in $\mathcal{T}$ we do not have to worry about that part now. What edges are incident to A? They are

$$\overline{AB}_3, \quad \overline{AC}_5, \quad \overline{AD}_9.$$

We now have to calculate the value $d(A, A) + weight(e)$ for each of these three edges to find out which is the minimum. (If two values are equal we simply choose the first of the two edges.) We have

$$d(A, A) + weight(\overline{AB}_3) = 0 + 3 = 3,$$
$$d(A, A) + weight(\overline{AC}_5) = 0 + 5 = 5,$$
$$d(A, A) + weight(\overline{AD}_9) = 0 + 9 = 9.$$

The minimum of these values is 3 so we choose $w = A$ and $e = \overline{AB}_3$. With this we can make the rest of the assignments in steps 3.3–5,

$$v \longleftarrow B,$$
$$\mathcal{T} \longleftarrow \{A\} \cup \{\overline{AB}_3, B\} = \{A, \overline{AB}_3, B\},$$
$$d(A, B) \longleftarrow d(A, A) + weight(\overline{AB}_3) = 3.$$

(b) Now we look for the first vertex w in $\mathcal{T}$ where $d(A, w) + weight(e)$ is minimized. There are now two possibilities for w, namely vertices A and B. We now look among the edges in $\mathcal{G} - \mathcal{T}$ that are incident to only one of the vertices A or B. We list the possible edges,

$$\overline{AC}_5, \quad \overline{AD}_9, \quad \overline{BD}_7, \quad \overline{BE}_{12}.$$

We now have to calculate $d(A, w) + weight(e)$ for each of these edges. Notice, when $w = A$ we use the edges incident to A but when $w = B$ we use the edges incident to B. We have

$$d(A, A) + weight(\overline{AC}_5) = 0 + 5 = 5,$$
$$d(A, A) + weight(\overline{AD}_9) = 0 + 9 = 9,$$
$$d(A, B) + weight(\overline{BD}_7) = 3 + 7 = 10,$$
$$d(A, B) + weight(\overline{BE}_{12}) = 3 + 12 = 12.$$

Recall, we had found $d(A, B) = 3$ in the last step. The minimum of these values is 5 so we choose $w = A$ and $e = \overline{AC}_5$. We now make the rest of the assignments in steps 3.3–5, writing them in a way that is more natural for us,

$$v = C,$$
$$\mathcal{T} = \{A, \overline{AB}_3, B, \overline{AC}_5, C\},$$
$$d(A, C) = d(A, A) + weight(\overline{AC}_5) = 0 + 5 = 5.$$

(c) Again we look for the first vertex w in $\mathcal{T}$ where $d(A, w) + weight(e)$ is minimized. There are now three possibilities for w, namely vertices A, B, and C. We look among the edges in $\mathcal{G} - \mathcal{T}$ that are incident to only one of the vertices A, B, or C. We list these edges,

$$\overline{AD}_9, \quad \overline{BD}_7, \quad \overline{BE}_{12}, \quad \overline{CD}_{11}, \quad \overline{CF}_2.$$

We now have to calculate $d(A, w) + weight(e)$ for each of these edges. Again, when $w = A$ we use the edges incident to A, when $w = B$ we use the edges incident to B, and when $w = C$ we use the edges incident to C. We have

$$d(A, A) + weight(\overline{AD}_9) = 0 + 9 = 9,$$
$$d(A, B) + weight(\overline{BD}_7) = 3 + 7 = 10,$$
$$d(A, B) + weight(\overline{BE}_{12}) = 3 + 12 = 15,$$
$$d(A, C) + weight(\overline{CD}_{11}) = 5 + 11 = 16,$$
$$d(A, C) + weight(\overline{CF}_2) = 5 + 2 = 7.$$

The minimum of these values is 7 so we choose $w = C$ and $e = \overline{CF}_2$. We now make the rest of the assignments in steps 3.3–5,

$$v = F,$$
$$\mathcal{T} = \{A, \overline{AB}_3, B, \overline{AC}_5, C, \overline{CF}_2, F\},$$
$$d(A, F) = d(A, C) + weight(\overline{CF}_2) = 5 + 2 = 7.$$

(d) Again we look for the first vertex w in $\mathcal{T}$ where $d(A, w) + weight(e)$ is minimized. There are now four possibilities for w, namely vertices A, B, C, and F. We look among the edges in $\mathcal{G} - \mathcal{T}$ that are incident to only one of these vertices. We list these edges,

$$\overline{AD}_9, \quad \overline{BD}_7, \quad \overline{BE}_{12}, \quad \overline{CD}_{11}, \quad \overline{DF}_8, \quad \overline{EF}_{10}.$$

We now have to calculate $d(A, w) + weight(e)$ for each of these edges. Again, when $w = A$ we use the edges incident to A, when $w = B$ we use the edges incident to B, and so on. We have

$$d(A, A) + weight(\overline{AD}_9) = 0 + 9 = 9,$$
$$d(A, B) + weight(\overline{BD}_7) = 3 + 7 = 10,$$
$$d(A, B) + weight(\overline{BE}_{12}) = 3 + 12 = 15,$$
$$d(A, C) + weight(\overline{CD}_{11}) = 5 + 11 = 16,$$
$$d(A, F) + weight(\overline{DF}_8) = 7 + 8 = 15,$$
$$d(A, F) + weight(\overline{EF}_{10}) = 7 + 10 = 17.$$

The minimum of these values is 9 so we choose $w = A$ and $e = \overline{AD}_9$. We now make the rest of the assignments in steps 3.3–5,

$$v = D,$$
$$\mathcal{T} = \{A, \overline{AB}_3, B, \overline{AC}_5, C, \overline{CF}_2, F, \overline{AD}_9, D\},$$
$$d(A, D) = d(A, A) + weight(\overline{AD}_9) = 0 + 9 = 9.$$

(e) Again we look for the first vertex w in $\mathcal{T}$ where $d(A, w) + weight(e)$ is minimized. There are now five possibilities for w, namely vertices A, B, C, F, and D. We look among the edges in $\mathcal{G} - \mathcal{T}$ that are incident to only one of these vertices. We list these edges,

$$\overline{BE}_{12}, \quad \overline{DE}_7, \quad \overline{EF}_{10}.$$

We now have to calculate $d(A, w) + weight(e)$ for each of these edges. Again, when $w = A$ we use the edges incident to A, when $w = B$ we use the edges incident to B, and so on. We have

$$d(A, B) + weight(\overline{BE}_{12}) = 3 + 12 = 15,$$
$$d(A, D) + weight(\overline{DE}_7) = 9 + 7 = 16,$$
$$d(A, F) + weight(\overline{EF}_{10}) = 7 + 10 = 17.$$

The minimum of these values is 15 so we choose $w = B$ and $e = \overline{BE}_{12}$. We now make the rest of the assignments in steps 3.3–5,

$$v = E,$$
$$\mathcal{T} = \{A, \overline{AB}_3, B, \overline{AC}_5, C, \overline{CF}_2, F, \overline{AD}_9, D, \overline{BE}_{12}, E\},$$
$$d(A, E) = d(A, B) + weight(\overline{BE}_{12}) = 3 + 12 = 15.$$

At this point we have gone through the **for-do** loop five times and we now move to step 4 which outputs the tree

$$\mathcal{T} = \{A, \overline{AB}_3, B, \overline{AC}_5, C, \overline{CF}_2, F, \overline{AD}_9, D, \overline{BE}_{12}, E\}.$$

See Fig. 10.3.

10.4 PROBLEMS

Question 10.1 *Suppose you have a tree $\mathcal{T}$ with 100 edges. How many vertices do you have? How many of the 100 edges are bridges?*

Question 10.2 *Suppose you have a tree $\mathcal{T}$ with 100 vertices. How many edges do you have? How many of the edges are bridges?*

Question 10.3 *Suppose you have a tree $\mathcal{T}$ with 100 edges. You remove two edges at random. Is the new graph disconnected? If so, how many components does the new graph have?*

Question 10.4 *Suppose you have a tree $\mathcal{T}$ with 20 vertices labeled $\{v_1, v_2, \ldots, v_{20}\}$. How many paths (with no repeated vertices or edges) are there between v_1 and v_{20}? How many paths are there between v_4 and v_{17}? How many paths are there between v_i and v_j for any i and j where $i \neq j$?*

Question 10.5 *Suppose you have a tree $\mathcal{T}$ with 20 vertices labeled $\{v_1, v_2, \ldots, v_{20}\}$. You add an edge between v_1 and v_{20}. Does the new graph have a cycle? If so, how many cycles does it have? Is the new graph still a tree?*

Question 10.6 *Suppose you have a rooted binary tree T. Suppose vertex v_i is not the root or a leaf. How many parents does v_i have? How many children does v_i have?*

Question 10.7 *Suppose you have a rooted strict binary tree T. Suppose vertex v_i is not the root or a leaf. How many parents does v_i have? How many children does v_i have?*

Question 10.8 *Given a tree T with 100 vertices, what is the greatest number of leaves that T can have? What is the smallest number of leaves that T can have?*

Question 10.9 *Suppose you have a graph with 6 vertices. One vertex has degree five and the other vertices have degree one. Draw the graph.*

Question 10.10 *For the graphs below answer the following questions.*
 (a) *How many spanning trees does the graph contain?*
 (b) *Which edge or edges must be contained in any spanning tree?*

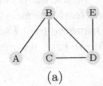
(a)

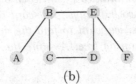
(b)

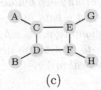
(c)

Question 10.11 *For the graphs below answer the following questions.*
 (a) *How many spanning trees does the graph contain?*
 (b) *Which edge or edges must be contained in any spanning tree?*

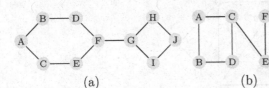
(a)   (b)

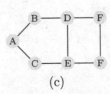
(c)

Question 10.12 *Use Prim's algorithm to find the minimal tree of the following graph starting at vertex A.*

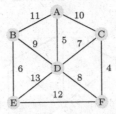

Question 10.13 *Use Prim's algorithm to find the minimal tree of the following graph starting at vertex A. Assume the edge set is given by $\mathcal{E} = \{\overline{AB}, \overline{AC}, \overline{BC}, \overline{CD}, \overline{CE}, \overline{CF}, \overline{DF}, \overline{EF}\}$.*

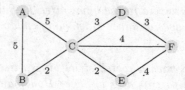

Question 10.14 *Suppose you have five buildings that you want to connect together using a local area network (LAN). A table giving the price (in thousands of dollars) to link each pair of building using fiber-optic cables is given below:*

	A	B	C	D	E
A	-				
B	24	-			
C	20	21	-		
D	28	25	18	-	
E	16	26	23	19	-

In order for any two buildings to communicate with each other over the LAN it is enough for there to exist a path between the two buildings. We want to find the least expensive possible LAN configuration. In other words, we want to find the minimal spanning tree. Use Prim's algorithm to find the minimal spanning tree. Assume the edges in the edge set are written in alphabetical order.

Question 10.15 *Use Dijkstra's algorithm to find the shortest path from vertex A to every other vertex. Assume the edge set is written in alphabetical order.*

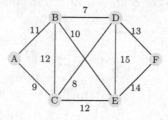

Question 10.16 *Use Dijkstra's algorithm to find the shortest path from vertex A to every other vertex.*

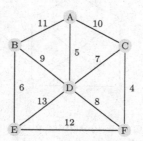

Basic Circuit Design

In this appendix we will try to see how several different topics interrelate in the context of computer science. This appendix is meant to give you a glimpse of things to come in computer science and to help illustrate in a concrete way how some of the topics we have already studied get used in the real world.

Computers use binary numbers because on a fundamental level computers can only operate on two states, the presence or absence of an electrical current. We humans interpret these signals as binary numbers. The most fundamental thing that a computer can do is the adding of two binary numbers. In this chapter we will see how computers do this.

A.1 BINARY ADDITION

Before seeing how the addition of binary numbers work we will remind ourselves of how addition works. We will restrict ourselves to adding only two numbers.

Example A.1

Add $82,376,514$ and $47,032,867$.

$$\begin{array}{r} \overset{1\ \ 1\ \ 1}{82\,376\,514} \\ +\ 47\,032\,867 \\ \hline 129\,409\,381 \end{array}$$

- In the first column (on the far right) add the two digits in that column, $4 + 7 = 11$. Write 1 below the line and carry 1.
- In the second column add the carried 1 along with the two digits in that column, $1 + 1 + 6 = 8$. Write 8 below the line.
- In the third column add the two digits in that column, $5 + 8 = 13$. Write 3 below the and carry 1.
- In the fourth column add the carried 1 along with the two digits in that column, $1 + 6 + 2 = 9$. Write 9 below the line.
- In the fifth column add the two digits in that column, $7 + 3 = 10$. Write 0 below the line and carry 1.
- In the sixth column add the carried 1 along with the two digits in that column, $1 + 3 + 0 = 4$. Write 4 below the line.
- In the seventh column add the digits in the column, $2 + 7 = 9$. Write 9 below the line.
- In the eighth column add the digits in the column, $8 + 4 = 12$. Here we write 12 below the line.

Something similar happens when we add binary numbers. First we write down all the sums of two single digit binary numbers;

$$0 + 0 = 0,$$
$$0 + 1 = 1,$$
$$1 + 0 = 1,$$
$$1 + 1 = 10.$$

Next, since dealing with carries requires us to add three single digit binary numbers we write down all these sums as well;

$$0 + 0 + 0 = 0, \qquad\qquad 1 + 0 + 0 = 1,$$
$$0 + 0 + 1 = 1, \qquad\qquad 1 + 0 + 1 = 10,$$
$$0 + 1 + 0 = 1, \qquad\qquad 1 + 1 + 0 = 10,$$
$$0 + 1 + 1 = 10, \qquad\qquad 1 + 1 + 1 = 11.$$

With this we are prepared to add any two binary numbers as we do in the next example.

Example A.2

Add the binary numbers 1100 1001 and 1110 1111. (Since it is clear we are dealing with binary numbers we will not place a small subscript of 2 after each binary number.)

$$\begin{array}{r} 1 \quad 1\,111 \\ 1100\ 1001 \\ +\ 1110\ 1111 \\ \hline 1\ 1011\ 1000 \end{array}$$

- In the first column (on the far right) add the two digits in that column, $1 + 1 = 10$. Write 0 below the line and carry 1.
- In the second column add the carried 1 along with the two digits in that column, $1 + 0 + 1 = 10$. Write 0 below the line and carry 1.
- In the third column add the carried 1 along with the two digits in that column, $1 + 0 + 1 = 10$. Write 0 below the line and carry 1.
- In the fourth column add the carried 1 along with the two digits in that column, $1 + 1 + 1 = 11$. Write 1 below the line and carry 1.
- In the fifth column add the carried 1 along with the two digits in that column, $1 + 0 + 0 = 1$. Write 1 below the line.
- In the sixth column add the two digits in that column, $0 + 1 = 1$. Write 1 below the line.
- In the seventh column add the two digits in that column, $1 + 1 = 10$. Write 0 below the line and carry 1.
- In the eighth column add the carried 1 along with the two digits in that column, $1 + 1 + 1 = 11$. Here we write 11 below the line.

In this example the digits that we write below the line are called **sum bits** and the digits that we carry (in gray) are called the **carry bits**.

A.2   THE HALF-ADDER

Our goal is to design a digital circuit using the logic gates presented in chapter 5 that adds two eight digit binary numbers. But first a little background. A single binary digit is called a **bit**, which got its name from **bi**nary dig**it** by combining the first two letter of binary with the last two letters of digit. Computers must store data in memory, and in order to both know where to put data and where to retrieve data, the areas of the computer's memory are given addresses. In most modern computers each address area consists of eight bits, which is called a **byte**. Thus, each addressable area of a computer's memory can contain an eight digit binary number. So a single byte can store the binary numbers 0000 0000 through 1111 1111, which are the decimal numbers 0 through 255. If a binary number is larger than 1111 1111 (as most are) then more than one byte must be used to store the number. This is not a course on computer architecture so we will not go into any more detail here, but this is enough to understand why we are interested in adding eight digit binary numbers.

Now we are ready to begin trying to design a circuit to add two eight digit binary numbers. First we again write down all the sums of two single digit binary numbers, only now we will write the answer using both a gray carry bit and a black sum bit;

$$0 + 0 = 00,$$
$$0 + 1 = 01,$$
$$1 + 0 = 01,$$
$$1 + 1 = 10.$$

As a first step we will design a circuit that does this addition. The key here is that we will split the addition into two parts; one part will give the sum bits and the other part will give the carry bits. We will rewrite the sum bits in a slightly different way that will hopefully seem familiar. Notice that we wrote the table in the reverse order of the sums.

x	y	sum bit
1	1	0
1	0	1
0	1	1
0	0	0

This should look very familiar, it is nothing more than the **xor** table from chapter 5.

x	y	x **xor** y
1	1	0
1	0	1
0	1	1
0	0	0

Thus the **xor** gate gives us the sum bit from this addition.

We now rewrite the gray carry bits in a table form. Again, we wrote the table in the reverse order of the sums.

x	y	carry bit
1	1	1
1	0	0
0	1	0
0	0	0

This is exactly the **and** table from chapter 5.

x	y	$x \times y$
1	1	1
1	0	0
0	1	0
0	0	0

Thus the **and** gate gives us the carry bit from this addition.

So if x is the first digit and y is the second digit then the **xor** gate gives us the sum bit and the **and** gate gives us the carry bit. The whole circuit is given by

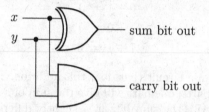

This circuit is called a **half-adder**. It works perfectly for adding two single-digit binary numbers. So, when it comes to adding two eight-digit binary numbers this circuit would work fine for the first column on the right where we do not yet have to worry about caries. But for the second column on, this circuit is not enough since it is possible we have carries. If we have a carry then we need to add three one-digit binary numbers.

A.3 THE FULL-ADDER

We want to be able to add a carry digit along with two other digits, $c + x + y$. In other words, we want to be able to add three single-digit binary numbers. We now rewrite sums of three single digit binary numbers using both a gray carry bit and a black sum bit;

$$0 + 0 + 0 = 00, \qquad 1 + 0 + 0 = 01,$$
$$0 + 0 + 1 = 01, \qquad 1 + 0 + 1 = 10,$$
$$0 + 1 + 0 = 01, \qquad 1 + 1 + 0 = 10,$$
$$0 + 1 + 1 = 10, \qquad 1 + 1 + 1 = 11.$$

This is what we want our circuit to do.

The basic idea is that we can do this in stages. First we add $x + y$ using the half-adder. This is what we get,

$$0 + 0 = 00,$$
$$0 + 1 = 01,$$
$$1 + 0 = 01,$$
$$1 + 1 = 10.$$

Then we take the sum digit that the half-adder outputs, shown in black above, and add it to the carry digit c using another half adder. The carry digit can either be $c = 1$ or $c = 0$. The addition of this carry digit with the sum digit turns out to be

$$0 + 0 = 00, \qquad\qquad 1 + 0 = 01,$$
$$0 + 1 = 01, \qquad\qquad 1 + 1 = 10,$$
$$0 + 1 = 01, \qquad\qquad 1 + 1 = 10,$$
$$0 + 0 = 00, \qquad\qquad 1 + 0 = 01.$$

Notice, the sum digits in black that we get here are exactly the same as the sum digits that we want to get when we add three single-digit binary numbers. So putting two half-adders in a row seems to give us the sum digits that we want. But what about the carry digits? These do not exactly match up with the carry digits we want when we add three numbers. Suppose

$$a = \text{Carry digit from the addition } x + y,$$
$$b = \text{Carry digit from the additon of } c \text{ and sum digit of } x + y.$$

Then we get the following table for $a \vee b$. (Note, in the chapter on Boolean Algebra we would have written this as $a + b$ but meant Boolean addition. Here since we are already talking about addition we will use the logical symbol for **or** to avoid confusion.)

a	b	$a \vee b$
0	0	0
0	0	0
0	0	0
1	0	1
0	0	0
0	1	1
0	1	1
1	0	1

And $a + b$ are exactly the carry digits we want from adding three digits. Thus, the circuit we want is a half-adder for $x + y$, and then another half-adder for the sum digit of $x + y$ and c, and an or gate for the carry digit of $x + y$ and c. In other words, we want this circuit:

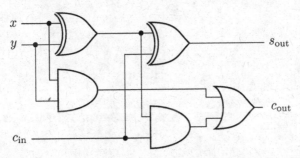

It would be a good idea to trace through this circuit for all of the possible values of x, y, and c_{in} and check that its outputs match the expected outputs of the additions from the start of the section.

A.4 ADDING TWO EIGHT-DIGIT BINARY NUMBERS

Our goal is to add two eight-digit binary numbers. We first found a circuit, called the half-adder, that would add two single-digit binary numbers $x + y$. We then used that to find a circuit, called a full-adder, that would add three single digit binary numbers $c + x + y$. Now we will find a circuit to add two eight-digit binary numbers. Suppose that we are adding $x_8 x_7 x_6 x_5 x_4 x_3 x_2 x_1$ and $y_8 y_7 y_6 y_5 y_4 y_3 y_2 y_1$, where each x_i and y_i is a single binary digit 0 or 1. Here we add these two numbers.

$$
\begin{array}{ccccccccc}
{}^{c_7} & {}^{c_6} & {}^{c_5} & {}^{c_4} & {}^{c_3} & {}^{c_2} & {}^{c_1} & {}^{0} & \\
x_8 & x_7 & x_6 & x_5 & x_4 & x_3 & x_2 & x_1 & \\
+\ y_8 & y_7 & y_6 & y_5 & y_4 & y_3 & y_2 & y_1 & \\
\hline
c_8\ s_8 & s_7 & s_6 & s_5 & s_4 & s_3 & s_2 & s_1 &
\end{array}
$$

In the first column (on the far right) we add $0 + x_1 + y_1$. This addition gives us a carry digit c_1 and a sum digit s_1. The sum digit s_1 is written below the line and the carry digit c_1 is carried to column two. In the second column we add $c_1 + x_2 + y_2$ to get a carry digit c_2 and a sum digit s_2. The sum digit s_2 is written below the line and the carry digit c_2 is carried to column three. Columns three through seven are similar. In column eight we add $c_7 + x_8 + y_8$ to give us both a carry digit c_8 and a sum digit s_8. Here we write both of these below the line. this is exactly what we want our circuit to do.

The way we do this is to link up eight full-adders in a row, taking the carry digit output from each full-adder as the carry digit input into the next full adder and recording the sum digit output from each full adder. This full circuit is shown in Fig. A.1. Notice, because we are using a full-adder for the addition $x_1 + y_1$, we automatically have the carry digit input into the first full-adder as zero.

If we wanted to add two sixteen-digit binary numbers then we would put two of these circuits in a row. and the carry bit output from the first circuit would be the carry bit input into the second circuit. And of course adding two twenty-four-digit binary numbers would require three of these circuits in row, and so on.

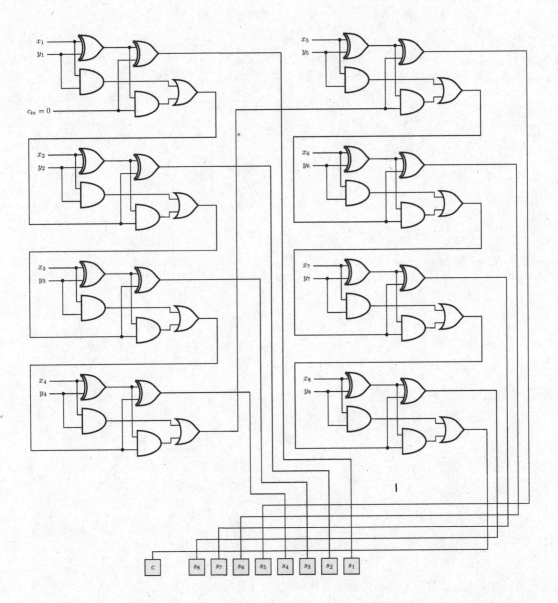

Figure A.1 The full circuit for an eight-bit adder. This circuit adds the two eight-big binary numbers $x_8x_7x_6x_5x_4x_3x_2x_1$ and $y_8y_7y_6y_5y_4y_3y_2y_1$ to obtain the binary number $cs_8s_7s_6s_5s_4s_3s_2s_1$. Two sixteen-bit binary numbers could be added together using two eight-bit adders in a row. The carry bit c would become the input c_{in} for the second eight-bit adder.

Answers to Problems

B.1   CHAPTER ONE ANSWERS

Question 1.1: This algorithm finds the number of seconds in a period of time that is given as a number of days, hours, minutes, and seconds. The output is $287,487$.

Question 1.2:

(a)

Step	x	y	max	Output
1	6	12	-	-
2.2	6	12	12	-
3	6	12	12	12

(c)

Step	x	y	max	Output
1	7	7	-	-
2.2	7	7	7	-
3	7	7	7	7

(b)

Step	x	y	max	Output
1	15	9	-	-
2.1	15	9	15	-
3	15	9	15	15

Question 1.3:

(a)

Step	$price$	$paid$	$change$	Output
1	74.36	82.50	-	-
2.2	74.36	82.50	8.84	-
3	74.36	82.50	8.84	Change 8.14

(b)

Step	$price$	$paid$	$change$	Output
1	75.00	55.50	-	-
2.1	75.00	55.50	-	Did not pay enough

(c)

Step	$price$	$paid$	$change$	Output
1	48.36	54.92	-	-
2.2	48.36	54.92	6.56	-
3	48.36	54.92	6.56	Change 6.56

Question 1.4:

(a)

Step	x	$answer$	Output
1	1	-	-
2	-1	-	-
3.1	-1	4	-
4	-1	4	4

(b)

Step	x	$answer$	Output
1	-3	-	-
2	3	-	-
3.2	3	-2	-
4	3	-2	-2

(c)

Step	x	answer	Output
1	7	-	-
2	-7	-	-
3.2	-7	-12	-
4	-7	-12	-12

Question 1.5:

(a)

Step	x	answer	Output
1	-6	-	-
2.1	-6	-9	-
3	-6	-9	-9

(b)

Step	x	answer	Output
1	2	-	-
2.2.2	2	-7	-
3	2	-7	-7

(c)

Step	x	answer	Output
1	8	-	-
2.2.1	8	-13	-
3	8	-13	-13

Question 1.6:

(a)

Step	x	n	i	answer	Output
1	2	4	-	-	-
2	2	4	-	2	-
3.1	2	4	1	4	-
3.1	2	4	2	8	-
3.1	2	4	3	16	-
4	2	4	3	16	16

(b)

Step	x	n	i	answer	Output
1	3	5	-	-	-
2	3	5	-	3	-
3.1	3	5	1	9	-
3.1	3	5	2	27	-
3.1	3	5	3	81	-
3.1	3	5	4	243	-
4	3	5	4	243	243

(c)

Step	x	n	i	answer	Output
1	5	6	-	-	-
2	5	6	-	5	-
3.1	5	6	1	25	-
3.1	5	6	2	125	-
3.1	5	6	3	625	-
3.1	5	6	4	3125	-
3.1	5	6	5	15625	-
4	5	6	5	15625	15625

Question 1.7:

(a)

Step	n	fac	Output
1	3	-	-
2	3	3	-
3.1	2	3	-
3.2	2	6	-
3.1	1	6	-
3.2	1	6	-
4	1	6	6

(b)

Step	n	fac	Output
1	5	-	-
2	5	5	-
3.1	4	5	-
3.2	4	20	-
3.1	3	20	-
3.2	3	60	-
3.1	2	60	-
3.2	2	120	-
3.1	1	120	-
3.2	1	120	-
4	1	120	120

(c)

Step	n	fac	Output
1	7	-	-
2	7	7	-
3.1	6	7	-
3.2	6	42	-
3.1	5	42	-
3.2	5	210	-
3.1	4	210	-
3.2	4	840	-
3.1	3	840	-
3.2	3	2520	-
3.1	2	2520	-
3.2	2	5040	-
3.1	1	5040	-
3.2	1	5040	-
4	1	5040	5040

Question 1.8:

(a)

Step	n	i	fac	Output
1	3	-	-	-
2	3	-	1	-
3.1	3	1	1	-
3.1	3	2	2	-
3.1	3	3	6	-
4	3	3	6	6

(b)

Step	n	i	fac	Output
1	5	-	-	-
2	5	-	1	-
3.1	5	1	1	-
3.1	5	2	2	-
3.1	5	3	6	-
3.1	5	4	24	-
3.1	5	5	120	-
4	5	5	120	120

(c)

Step	n	i	fac	Output
1	7	-	-	-
2	7	-	1	-
3.1	7	1	1	-
3.1	7	2	2	-
3.1	7	3	6	-
3.1	7	4	24	-
3.1	7	5	120	-
3.1	7	6	720	-
3.1	7	7	5040	-
4	7	7	5040	5040

Question 1.9:

Step	d	n	Output
1	-	$8, 490, 725, 727, 154, 368, 726, 402, 945$	-
2	25	$8, 490, 725, 727, 154, 368, 726, 402, 945$	-
3.1	25	$117 (\leftarrow 8 + 4 + 9 + 0 + 7 + \cdots + 4 + 5$	-
3.2	3	117	-
3.1	3	$9 (\leftarrow 1 + 1 + 7)$	-
4	1	9	9

Question 1.10:

(a) $x_1 = 2$, $x_2 = 7$, $x_3 = 4$, $x_4 = 3$, $x_5 = 9$, $x_6 = 0$, $s = 9$

Step	i	x_i	s	Output
2.1	1	2	9	-
2.1	2	7	9	-
2.1	3	4	9	-
2.1	4	3	9	-
2.1	5	9	9	-
2.1.1	5	9	9	String contains 9.

(b) $x_1 = 7$, $x_2 = 3$, $x_3 = 0$, $x_4 = 2$, $x_5 = 8$, $x_6 = 5$, $s = 2$

Step	i	x_i	s	Output
2.1	1	7	2	-
2.1	2	3	2	-
2.1	3	0	2	-
2.1	4	2	2	-
2.1.1	4	2	2	String contains 2.

(c) $x_1 = 9$, $x_2 = 8$, $x_3 = 3$, $x_4 = 6$, $x_5 = 5$, $x_6 = 0$, $x_7 = 3$, $x_8 = 6$, $x_9 = 2$, $s = 5$

Step	i	x_i	s	Output
2.1	1	9	5	-
2.1	2	8	5	-
2.1	3	3	5	-
2.1	4	6	5	-
2.1	5	5	5	-
2.1.1	5	5	5	String contains 5.

Question 1.11:

(a) $x_1 = 2$, $x_2 = 1$, $x_3 = 0$, $x_4 = 9$, $y_1 = 4$, $y_2 = 0$, $y_3 = 7$, $y_4 = 1$, $s = 0$

Step	i	x_i	y_i	Output
2.1	1	2	-	-
2.1	2	1	-	-
2.1	3	0	-	-
2.1.1	3	0	-	String one contains 0.

(b) $x_1 = 5$, $x_2 = 9$, $x_3 = 3$, $y_1 = 7$, $y_2 = 2$, $y_3 = 1$, $s = 2$

Step	i	x_i	y_i	Output
2.1	1	5	-	-
2.1	2	9	-	-
2.1	3	3	-	-
3.1	1	-	7	-
3.1	2	-	2	-
3.1.1	2	-	2	String two contains 2.

(c) $x_1 = 3$, $x_2 = 9$, $x_3 = 7$, $y_1 = 4$, $y_2 = 4$, $y_3 = 3$, $s = 6$

Step	i	x_i	y_i	Output
2.1	1	3	-	-
2.1	2	9	-	-
2.1	3	7	-	-
3.1	1	-	4	-
3.1	2	-	4	-
3.1	3	-	7	-
4	3	-	-	Neither string contains 6.

Question 1.12:

(a) $x_1 = 2$, $x_2 = 7$, $x_3 = 3$, $y_1 = 4$, $y_2 = 3$, $y_3 = 9$, $n = 3$

Step	i	j	x_i	y_j	Output
2.1.1	1	1	2	4	-
2.1.1	1	2	2	3	-
2.1.1	1	3	2	9	-
2.1.1	2	1	7	4	-
2.1.1	2	2	7	3	-
2.1.1	2	3	7	9	-
2.1.1	3	1	3	4	-
2.1.1	3	2	3	3	-
2.1.1.1	3	2	3	3	There is an element common to both strings.

(b) $x_1 = 5$, $x_2 = 9$, $x_3 = 3$, $y_1 = 7$, $y_2 = 2$, $y_3 = 5$, $n = 3$

Step	i	j	x_i	y_j	Output
2.1.1	1	1	5	7	-
2.1.1	1	2	5	2	-
2.1.1	1	3	5	5	-
2.1.1	1	3	5	5	There is an element common to both strings.

(c) $x_1 = 1$, $x_2 = 2$, $x_3 = 3$, $y_1 = 9$, $y_2 = 8$, $y_3 = 7$, $n = 3$

Step	i	j	x_i	y_j	Output
2.1.1	1	1	1	9	-
2.1.1	1	2	1	8	-
2.1.1	1	3	1	7	-
2.1.1	2	1	2	9	-
2.1.1	2	2	2	8	-
2.1.1	2	3	2	7	-
2.1.1	3	1	3	9	-
2.1.1	3	2	3	8	-
2.1.1	3	3	3	7	-
3	3	3	3	7	There is no element common to both strings.

Question 1.13:

(a) $x_1 = 6$, $x_2 = 3$, $x_3 = 5$, $x_4 = 2$, $x_5 = 4$

Step	i	j	x_i	x_j	Output
2.1.1	1	2	6	3	-
2.1.1	1	3	6	5	-
2.1.1	1	4	6	2	-
2.1.1	1	5	6	4	-
2.1.1	2	3	3	5	-
2.1.1	2	4	3	2	-
2.1.1	2	5	3	4	-
2.1.1	3	4	5	2	-
2.1.1	3	5	5	4	-
2.1.1	4	5	2	4	-
3	4	5	2	4	There are no duplicate integer in the string.

(b) $x_1 = 4$, $x_2 = 7$, $x_3 = 3$, $x_4 = 7$, $x_5 = 2$, $x_6 = 9$, $x_7 = 1$, $x_8 = 6$

Step	i	j	x_i	x_j	Output
2.1.1	1	2	4	7	-
2.1.1	1	3	4	3	-
2.1.1	1	4	4	7	-
2.1.1	1	5	4	2	-
2.1.1	1	6	4	9	-
2.1.1	1	7	4	1	-
2.1.1	1	8	4	6	-
2.1.1	2	3	7	3	-
2.1.1	2	4	7	7	-
2.1.1.1	2	4	7	7	There is a duplicate integer in the string.

(c) $x_1 = 5$, $x_2 = 6$, $x_3 = 3$, $x_4 = 7$, $x_5 = 3$, $x_6 = 2$

Step	i	j	x_i	x_j	Output
2.1.1	1	2	5	6	-
2.1.1	1	3	5	3	-
2.1.1	1	4	5	7	-
2.1.1	1	5	5	3	-
2.1.1	1	6	5	1	-
2.1.1	2	3	6	3	-
2.1.1	2	4	6	7	-
2.1.1	2	5	6	3	-
2.1.1	2	6	6	2	-
2.1.1	3	4	3	7	-
2.1.1	3	5	3	3	-
2.1.1.1	3	5	3	3	There is a duplicate integer in the string.

Question 1.14:

(a)

Step	i	sum	Output
2	-	0	-
3.1	1	1	-
3.1	2	5	-
3.1	3	14	-
4	3	14	14

(b)

Step	i	sum	Output
2	-	0	-
3.1	1	1	-
3.1	2	5	-
3.1	3	14	-
3.1	4	30	-
4	4	30	30

(c)

Step	i	sum	Output
2	-	0	-
3.1	1	1	-
3.1	2	5	-
3.1	3	14	-
3.1	4	30	-
3.1	5	55	-
3.1	6	91	-
4	6	91	91

Question 1.15:

(a)

Step	n	$order$	Output
1	12	-	-
2	12	0	-
3.1	6	0	-
3.2	6	1	-
3.1	3	1	-
3.2	3	2	-
4	3	2	2

(b)

Step	n	$order$	Output
1	90	-	-
2	90	0	-
3.1	45	0	-
3.2	45	1	-
4	45	1	1

(c)

Step	n	$order$	Output
1	80	-	-
2	80	0	-
3.1	40	0	-
3.2	40	1	-
3.1	20	1	-
3.2	20	2	-
3.1	10	2	-
3.2	10	3	-
3.1	5	3	-
3.2	5	4	-
4	5	4	4

Question 1.16:

(a) $x_1 = 5$, $x_2 = 7$, $x_3 = 4$, $x_4 = 6$, $x_5 = 2$, $x_6 = 8$

Step	min	position	i	x_i	Output
2	5	1	-	-	-
3.1	5	1	2	7	-
3.1	5	1	3	4	-
3.1.1	4	1	3	4	-
3.1.2	4	3	3	4	-
3.1	4	3	4	6	-
3.1	4	3	5	2	-
3.1.1	2	3	5	2	-
3.1.2	2	5	5	2	-
3.1	2	5	6	8	-
4	2	5	6	8	2,5

(b) $x_1 = 3$, $x_2 = 5$, $x_3 = 7$, $x_4 = 2$, $x_5 = 4$, $x_6 = 6$

Step	min	position	i	x_i	Output
2	3	1	-	-	-
3.1	3	1	2	5	-
3.1	3	1	3	7	-
3.1	3	1	4	2	-
3.1.1	2	1	4	2	-
3.1.2	2	4	4	2	-
3.1	2	4	5	4	-
3.1	2	4	6	6	-
4	2	4	6	6	2,4

(c) $x_1 = 7$, $x_2 = 2$, $x_3 = 5$, $x_4 = 0$, $x_5 = 9$, $x_6 = 3$, $x_7 = 1$

Step	min	position	i	x_i	Output
2	7	1	-	-	-
3.1	7	1	2	2	-
3.1.1	2	1	2	2	-
3.1.2	2	2	2	2	-
3.1	2	2	3	5	-
3.1	2	2	4	0	-
3.1.1	0	2	4	0	-
3.1.2	0	4	4	0	-
3.1	0	4	5	9	-
3.1	0	4	6	3	-
3.1	0	4	7	1	-
4	0	4	7	1	0,4

Question 1.17:

(a) $x_1 = 0$, $x_2 = 4$, $x_3 = m$, $x_4 = 3$, $x_5 = s$, $x_7 = 8$

Step	i	x_i	noninteger-detected	Output
2	1	-	-	-
3	1	-	false	-
4.1	1	0	false	-
4.2	2	0	false	-
4.1	2	4	false	-
4.2	3	4	false	-
4.1	3	m	false	-
4.1.1	3	m	true	-
4.1	4	m	true	-
5.1	4	m	true	String contains non-integer character.

(b) $x_1 = 8$, $x_2 = 9$, $x_3 = 0$

Step	i	x_i	noninteger-detected	Output
2	1	-	-	-
3	1	-	false	-
4.1	1	8	false	-
4.2	2	8	false	-
4.1	2	9	false	-
4.2	3	9	false	-
4.1	3	0	false	-
5.2	3	0	false	String consists entirely of integers.

(c) $x_1 = 7$, $x_2 = 3$, $x_3 = 2$, $x_4 = r$, $x_5 = 9$

Step	i	x_i	noninteger_ detected	Output
2	1	-	-	-
3	1	-	false	-
4.1	1	7	false	-
4.2	2	7	false	-
4.1	2	3	false	-
4.2	3	3	false	-
4.1	3	2	false	-
4.2	4	2	false	-
4.1	4	r	false	-
4.1.1	4	r	true	-
4.2	5	r	true	-
5.1	5	r	true	String contains non-digit character.

Question 1.18: One possible solution:
1. Input string $x_1, x_2, \ldots, x_n$.
2. **For** $i = 1$ to n **do**
    2.1. **If** x_i is not a digit **then**
        2.1.1. Return "String contains non-digit characters."
3. Return "String consists entirely of digits."

B.2   CHAPTER TWO ANSWERS

Question 2.1:

(a) $101_2 = 5_{10}$
(b) $110_2 = 6_{10}$
(c) $111_2 = 7_{10}$
(d) $1011_2 = 11_{10}$
(e) $1101_2 = 13_{10}$

(f) $1010_2 = 10_{10}$
(g) $10010_2 = 18_{10}$
(h) $10110_2 = 22_{10}$
(i) $11111_2 = 31_{10}$

Question 2.2:

(a) $0110\ 1001_2 = 105_{10}$
(b) $1011\ 1100_2 = 188_{10}$
(c) $0011\ 0111_2 = 55_{10}$
(d) $1111\ 0101_2 = 245_{10}$
(e) $1001\ 0110_2 = 150_{10}$

(f) $0110\ 1000_2 = 104_{10}$
(g) $1011\ 0110_2 = 182_{10}$
(h) $0010\ 1011_2 = 43_{10}$
(i) $1101\ 1101_2 = 221_{10}$

Question 2.3:

(a) $27_8 = 23_{10}$
(b) $42_8 = 34_{10}$
(c) $35_8 = 29_{10}$
(d) $103_8 = 67_{10}$
(e) $673_8 = 443_{10}$

(f) $360_8 = 240_{10}$
(g) $4721_8 = 2513_{10}$
(h) $2451_8 = 1321_{10}$
(i) $7715_8 = 4045_{10}$

Question 2.4:
  (a) $7254_8 = 3756_{10}$
  (b) $1602_8 = 898_{10}$
  (c) $4640_8 = 2464_{10}$
  (d) $17530_8 = 8024_{10}$
  (e) $72501_8 = 30017_{10}$

  (f) $11101_8 = 4673_{10}$
  (g) $365310_8 = 125640_{10}$
  (h) $772531_8 = 259417_{10}$
  (i) $417524_8 = 139092_{10}$

Question 2.5:
  (a) $A3_{16} = 163_{10}$
  (b) $9F_{16} = 159_{10}$
  (c) $17_{16} = 23_{10}$
  (d) $3D2_{16} = 978_{10}$
  (e) $9A0_{16} = 2464_{10}$

  (f) $AED_{16} = 2797_{10}$
  (g) $F001_{16} = 61441_{10}$
  (h) $6D27_{16} = 27943_{10}$
  (i) $39CB_{16} = 14795_{10}$

Question 2.6:
  (a) $802_{16} = 2050_{10}$
  (b) $6A3_{16} = 1699_{10}$
  (c) $F2E_{16} = 3886_{10}$
  (d) $290B_{16} = 10507_{10}$
  (e) $4C71_{16} = 19569_{10}$

  (f) $1101_{16} = 4353_{10}$
  (g) $ABCDE_{16} = 703710_{10}$
  (h) $2E916_{16} = 190742_{10}$
  (i) $97CA0_{16} = 621728_{10}$

Question 2.7:
  (a) $0.11_2 = 0.75_{10}$
  (b) $0.01_2 = 0.25_{10}$
  (c) $0.10_2 = 0.5_{10}$
  (d) $0.110_2 = 0.75_{10}$
  (e) $0.101_2 = 0.625_{10}$

  (f) $0.011_2 = 0.375_{10}$
  (g) $0.0111_2 = 0.4375_{10}$
  (h) $0.0011_2 = 0.1875_{10}$
  (i) $0.1001_2 = 0.5625_{10}$

Question 2.8:
  (a) $100.110_2 = 4.75_{10}$
  (b) $110.011_2 = 6.375_{10}$
  (c) $111.001_2 = 7.125_{10}$
  (d) $1101.1000_2 = 13.5_{10}$
  (e) $0110.0111_2 = 6.4375_{10}$

  (f) $1011.1001_2 = 11.5625_{10}$
  (g) $1111.1111_2 = 15.9375_{10}$
  (h) $0001.0001_2 = 1.0625_{10}$
  (i) $1100.0101_2 = 12.3125_{10}$

Question 2.9:
  (a) $0.4_8 = 0.5_{10}$
  (b) $0.1_8 = 0.125_{10}$
  (c) $0.7_8 = 0.875_{10}$
  (d) $0.72_8 = 0.90625_{10}$
  (e) $0.03_8 = 0.046875_{10}$

  (f) $0.14_8 = 0.1875_{10}$
  (g) $0.553_8 = 0.708984375_{10}$
  (h) $0.407_8 = 0.513671875_{10}$
  (i) $0.321_8 = 0.408203125_{10}$

Question 2.10:
  (a) $7.5_8 = 7.625_{10}$
  (b) $4.2_8 = 4.25_{10}$
  (c) $3.6_8 = 3.75_{10}$
  (d) $37.63_8 = 31.796875_{10}$
  (e) $56.12_8 = 46.15625_{10}$

  (f) $25.25_8 = 21.328125_{10}$
  (g) $427.014_8 = 279.0234375_{10}$
  (h) $510.442_8 = 328.56640625_{10}$
  (i) $703.635_8 = 451.806640625_{10}$

Question 2.11:

(a) $0.7_{16} = 0.4375_{10}$

(b) $0.A_{16} = 0.625_{10}$

(c) $0.3_{16} = 0.1875_{10}$

(d) $0.A8_{16} = 0.65625_{10}$

(e) $0.4C_{16} = 0.295875_{10}$

(f) $0.82_{16} = 0.5078125_{10}$

(g) $0.C48_{16} = 0.767578125_{10}$

(h) $0.379_{16} = 0.2170410156_{10}$

(i) $0.ABC_{16} = 0.6708984375_{10}$

Question 2.12:

(a) $9.3_{16} = 9.1875_{10}$

(b) $B.E_{16} = 11.875_{10}$

(c) $4.C_{16} = 4.75_{10}$

(d) $AC.DC_{16} = 172.859375_{10}$

(e) $F2.39_{16} = 242.22265625_{10}$

(f) $5E.C5_{16} = 94.76953125_{10}$

(g) $A04.BB8_{16} = 2564.732421875_{10}$

(h) $7CF.9F0_{16} = 1999.62109375_{10}$

(i) $101.101_{16} = 257.06274414063_{10}$

Question 2.13:

(a) $62_8 = 110010_2$

(b) $31_8 = 11001_2$

(c) $571_8 = 101111001_2$

(d) $31.72_8 = 11001.11101_2$

(e) $42.15_8 = 100010.001101_2$

(f) $72.37_8 = 111010.011111_2$

(g) $313.011_8 = 11001011.000001001_2$

(h) $643.026_8 = 110100011.00001011_2$

(i) $211.361_8 = 10001001.011110001_2$

Question 2.14:

(a) $A0_{16} = 10100000_2$

(b) $3B_{16} = 111011_2$

(c) $27_{16} = 100111_2$

(d) $F2.F2_{16} = 11110010.1111001_2$

(e) $5B.93_{16} = 1011011.10010011_2$

(f) $88.7C_{16} = 10001000.011111_2$

(g) $214.3C5_{16} = 1000010100.0011111000101_2$

(h) $AD2.0BC_{16} = 101011010010.0000101111_2$

(i) $101.011_{16} = 100000001.000000010001_2$

Question 2.15:

(a) $10_2 = 2_8$

(b) $110_2 = 6_8$

(c) $11_2 = 3_8$

(d) $110.111_2 = 6.7_8$

(e) $011.010_2 = 3.2_8$

(f) $101.001_2 = 5.1_8$

(g) $11011.001001_2 = 33.11_8$

(h) $100011.10111001_2 = 43.562_8$

(i) $11110001.00011101_2 = 361.072_8$

Question 2.16:

(a) $1101_2 = D_{16}$

(b) $10_2 = 2_{16}$

(c) $101_2 = 5_{16}$

(d) $1110.1101_2 = E.D_{16}$

(e) $1011.1010_2 = B.A_{16}$

(f) $0101.0011_2 = 5.3_{16}$

(g) $111001011.0011011_2 = 1CB.36_{16}$

(h) $1011101.01011010101_2 = 5D.5AA_{16}$

(i) $10111100101.100111_2 = 5E5.9C_{16}$

Question 2.17:

(a) $7_{10} = 111_2$

(b) $3_{10} = 11_2$

(c) $6_{10} = 110_2$

(d) $28_{10} = 11100_2$

(e) $39_{10} = 100111_2$

(f) $79_{10} = 1001111_2$

(g) $381_{10} = 101111101_2$

(h) $643_{10} = 1010000011_2$

(i) $569_{10} = 1000111001_2$

Question 2.18:

(a) $78_{10} = 116_8$

(b) $93_{10} = 135_8$

(c) $52_{10} = 64_8$

(d) $295_{10} = 447_8$

(e) $944_{10} = 1660_8$

(f) $641_{10} = 1201_8$

(g) $4862_{10} = 11376_8$

(h) $5078_{10} = 11726_8$

(i) $1532_{10} = 2774_8$

Question 2.19:

(a) $34_{10} = 22_{16}$

(b) $23_{10} = 17_{16}$

(c) $536_{10} = 218_{16}$

(d) $285_{10} = 11D_{16}$

(e) $1897_{10} = 769_{16}$

(f) $8321_{10} = 2081_{16}$

(g) $90867_{10} = 162F3_{16}$

(h) $42778_{10} = A71A_{16}$

(i) $28714_{10} = 702A_{16}$

Question 2.20:

(a) $0.5_{10} = 0.1_2$

(b) $0.4_{10} \approx 0.011001_2$

(c) $0.8_{10} \approx 0.110011_2$

(d) $0.32_{10} \approx 0.010100_2$

(e) $0.81_{10} \approx 0.110011_2$

(f) $0.77_{10} \approx 0.110001_2$

(g) $0.239_{10} \approx 0.001111_2$

(h) $0.552_{10} \approx 0.100011_2$

(i) $0.798_{10} \approx 0.110011_2$

Question 2.21:

(a) $0.9_{10} \approx 0.714631_8$

(b) $0.4_{10} \approx 0.314631_8$

(c) $0.6_{10} \approx 0.463146_8$

(d) $0.83_{10} \approx 0.650753_8$

(e) $0.25_{10} \approx 0.2_8$

(f) $0.44_{10} \approx 0.341217_8$

(g) $0.482_{10} \approx 0.366621_8$

(h) $0.667_{10} \approx 0.525402_8$

(i) $0.315_{10} \approx 0.241217_8$

Question 2.22:

(a) $0.1_{10} \approx 0.199999_{16}$

(b) $0.3_{10} \approx 0.4CCCCC_{16}$

(c) $0.9_{10} \approx 0.E66666_{16}$

(d) $0.11_{10} \approx 0.1C28F5_{16}$

(e) $0.83_{10} \approx 0.D47AE1_{16}$

(f) $0.47_{10} \approx 0.7851EB_{16}$

(g) $0.429_{10} \approx 0.6DD2F1_{16}$

(h) $0.638_{10} \approx 0.A353F7_{16}$

(i) $0.314_{10} \approx 0.50624D_{16}$

B.3   CHAPTER THREE ANSWERS

Question 3.1:

(a) No, an opinion.

(b) No, a question.

(c) Yes.

(d) Yes.

(e) Yes.

(f) Yes.

(g) No, a request.

(h) No, a request.

(i) Yes.

Question 3.2:

(a) Jon plays baseball or Ron plays football.

(b) Ron plays football and Jon plays baseball.

(c) Jon does not play baseball and Ron plays football.

(d) Jon plays baseball and Ron does not play football.

(e) Ron does not play football or Jon does not play baseball.

(f) It is not the case that Ron plays football or Jon plays baseball.

Question 3.3:

(a) $p \wedge q \equiv T \wedge F \equiv F$

(b) $q \wedge p \equiv F \wedge T \equiv F$

(c) $\neg q \vee p \equiv \neg F \vee T \equiv T \vee T \equiv T$

(d) $\neg p \vee q \equiv \neg T \vee F \equiv F \vee F \equiv F$

(e) $p \wedge \neg q \equiv T \wedge \neg F \equiv T \wedge T \equiv T$

(f) $\neg q \vee \neg p \equiv \neg F \vee \neg T \equiv T \vee F \equiv T$

(g) $p \rightarrow q \equiv T \rightarrow F \equiv F$

(h) $\neg q \rightarrow p \equiv \neg F \rightarrow T \equiv T \rightarrow T \equiv T$

(i) $\neg q \rightarrow \neg p \equiv \neg F \rightarrow \neg T \equiv T \rightarrow F \equiv F$

(j) $p \leftrightarrow q \equiv T \leftrightarrow F \equiv F$

(k) $\neg q \leftrightarrow p \equiv \neg F \leftrightarrow T \equiv T \leftrightarrow T \equiv T$

(l) $\neg p \leftrightarrow \neg q \equiv \neg T \leftrightarrow \neg F \equiv F \leftrightarrow T \equiv F$

Question 3.4:

(a)

p	q	$p \wedge q$
T	T	T
T	F	F
F	T	F
F	F	F

(d)

p	q	$\neg p \vee q$
T	T	T
T	F	F
F	T	T
F	F	T

(g)

p	q	$p \rightarrow q$
T	T	T
T	F	F
F	T	T
F	F	T

(j)

p	q	$p \leftrightarrow q$
T	T	T
T	F	F
F	T	F
F	F	T

(b)

p	q	$q \wedge p$
T	T	T
T	F	F
F	T	F
F	F	F

(e)

p	q	$p \wedge \neg q$
T	T	F
T	F	T
F	T	F
F	F	F

(h)

p	q	$\neg q \rightarrow p$
T	T	T
T	F	T
F	T	T
F	F	F

(k)

p	q	$\neg q \leftrightarrow p$
T	T	F
T	F	T
F	T	T
F	F	F

(c)

p	q	$\neg q \vee p$
T	T	T
T	F	T
F	T	F
F	F	T

(f)

p	q	$\neg q \vee \neg p$
T	T	F
T	F	T
F	T	T
F	F	T

(i)

p	q	$\neg q \rightarrow \neg p$
T	T	T
T	F	F
F	T	T
F	F	T

(l)

p	q	$\neg p \leftrightarrow \neg q$
T	T	T
T	F	F
F	T	F
F	F	T

Question 3.5:

We have

p	q	$p \rightarrow q$
T	T	T
T	F	F
F	T	T
F	F	T

and

p	q	$\neg p \vee q$
T	T	T
T	F	F
F	T	T
F	F	T

.

Since both $p \rightarrow q$ and $\neg p \vee q$ have identical truth tables we can say that $p \rightarrow q \equiv \neg p \vee q$.

Question 3.6:

We have

p	q	$p \leftrightarrow q$
T	T	T
T	F	F
F	T	F
F	F	T

and

p	q	$p \rightarrow q$	$q \rightarrow p$	$(p \rightarrow q) \wedge (q \rightarrow p)$
T	T	T	T	T
T	F	F	T	F
F	T	T	F	F
F	F	T	T	T

.

Since both $p \leftrightarrow q$ and $(p \rightarrow q) \wedge (q \rightarrow p)$ have identical truth tables we can say they are equivalent. Using the implication law we can write $p \leftrightarrow q \equiv (\neg p \vee q) \wedge (\neg q \vee p)$.

Question 3.7:

(a)

p	q	r	$(p \vee \neg r) \to q$
T	T	T	T
T	T	F	T
T	F	T	F
T	F	F	F
F	T	T	T
F	T	F	T
F	F	T	T
F	F	F	F

(b)

p	q	$(q \wedge p) \to \neg p$
T	T	F
T	F	T
F	T	T
F	F	T

(c)

p	q	$\neg p \to (\neg p \vee q)$
T	T	T
T	F	T
F	T	T
F	F	T

(d)

p	q	r	$\neg(p \wedge \neg q) \vee r$
T	T	T	T
T	T	F	T
T	F	T	T
T	F	F	F
F	T	T	T
F	T	F	T
F	F	T	T
F	F	F	T

(e)

p	q	r	$\neg[(\neg p \wedge q) \vee r]$
T	T	T	F
T	T	F	T
T	F	T	F
T	F	F	T
F	T	T	F
F	T	F	F
F	F	T	F
F	F	F	T

(f)

p	q	r	$(p \wedge \neg q) \wedge (r \vee q)$
T	T	T	F
T	T	F	F
T	F	T	T
T	F	F	F
F	T	T	F
F	T	F	F
F	F	T	F
F	F	F	F

(g)

p	q	r	$(p \vee q) \leftrightarrow [p \vee (q \wedge r)]$
T	T	T	T
T	T	F	T
T	F	T	T
T	F	F	T
F	T	T	T
F	T	F	F
F	F	T	T
F	F	F	T

(h)

p	q	r	$(q \wedge \neg r) \leftrightarrow \neg(p \vee r)$
T	T	T	T
T	T	F	F
T	F	T	T
T	F	F	T
F	T	T	T
F	T	F	T
F	F	T	T
F	F	F	F

(i)

p	q	r	$(p \leftrightarrow q) \wedge (r \leftrightarrow q)$
T	T	T	T
T	T	F	F
T	F	T	F
T	F	F	F
F	T	T	F
F	T	F	F
F	F	T	F
F	F	F	T

Question 3.8:

(a) $p \vee p = p$

(b) $p \vee \neg p = T$

(c) $\neg(p \vee \neg p) = F$

(d) $p \vee (p \wedge q) = p$

(e) $p \vee (\neg p \wedge q) = p \vee q$

(f) $p \wedge (\neg p \vee q) = p \wedge q$

(g) $p \wedge p = p$

(h) $p \wedge (p \vee q) = p$

(i) $p \wedge (p \vee q \vee r) = p$

(j) $(p \wedge q) \vee (p \wedge \neg q) = p$

(k) $\neg(\neg p \vee \neg p) = p$

(l) $(p \wedge q) \vee (\neg p \wedge q) = q$

Question 3.9:

(a) $(\neg p \vee \neg q) \wedge (\neg p \vee q) = \neg p$

(b) $q \vee (q \wedge \neg q) = q$

(c) $\neg p \vee (q \wedge \neg p) = \neg p$

(d) $(p \vee \neg q) \wedge (p \vee q) = p$

(e) $r \vee [r \vee (r \wedge p)] = r$

(f) $p \wedge [p \vee (p \wedge q)] = p$

(g) $r \vee (r \wedge \neg p \wedge q \wedge s) = r$

(h) $\neg r \wedge \neg (r \wedge p \wedge q \wedge s) = \neg r$

Question 3.10:

(a) $q \wedge [p \vee (\neg p \wedge q)] = q$

(b) $\neg[(p \wedge \neg q) \vee \neg p] = p \wedge q$

(c) $(\neg p \wedge q) \vee [p \wedge (p \vee q)] = p \vee q$

(d) $\neg[p \vee (p \wedge q)] \wedge q = \neg p \wedge q$

(e) $(p \wedge \neg r) \vee (\neg p \wedge q) \vee \neg (q \wedge r) = \neg p \vee \neg q \vee \neg r$

(f) $\neg(p \wedge q) \vee (q \wedge r) = \neg p \vee \neg q \vee r$

(g) $[p \vee (q \wedge r)] \wedge (\neg p \vee r) = (p \vee q) \wedge r$

(h) $(p \wedge \neg r) \vee (\neg p \wedge q) \vee \neg (p \wedge r) = \neg p \vee \neg r$

(i) $(p \wedge r) \vee (\neg p \wedge q) \vee (r \wedge q) = (p \wedge r) \vee (q \wedge \neg p)$

(j) $\neg p \vee \neg q \vee [p \wedge q \wedge \neg r] = \neg p \vee \neg q \vee \neg r$

(k) $(p \vee r) \wedge (\neg p \vee q) \wedge (r \vee q) = (p \wedge q) \vee (r \wedge \neg p)$

(l) $\neg(p \wedge q) \wedge (\neg p \vee q) \wedge (\neg q \vee q) = \neg p$

Question 3.11:

(a) tautology

(b) not a tautology

(c) tautology

(d) not a tautology

Question 3.12

(a) contradiction

(b) not a contradiction

(c) not a contradiction

(d) contradiction

B.4   CHAPTER FOUR ANSWERS

Question 4.1:

(a) True	(g) False	(m) False	(s) True
(b) True	(h) True	(n) False	(t) False
(c) False	(i) True	(o) False	(u) True
(d) True	(j) False	(p) True	(v) False
(e) False	(k) True	(q) True	(w) True
(f) False	(l) False	(r) True	(x) False

Question 4.2:

(a) True	(g) False	(m) False	(s) False
(b) True	(h) True	(n) True	(t) True
(c) False	(i) True	(o) False	(u) False
(d) False	(j) True	(p) False	(v) False
(e) True	(k) False	(q) False	(w) False
(f) True	(l) True	(r) True	(x) True

Question 4.3:

(a) True	(f) True	(k) True	(p) True
(b) True	(g) False	(l) True	(q) True
(c) False	(h) True	(m) True	(r) True
(d) True	(i) True	(n) True	(s) False
(e) False	(j) True	(o) False	(t) True

Question 4.4:
 (a) $\{\ \}, \{a\}$
 (b) $\{\ \}, \{5\}$
 (c) $\{\ \}, \{v\}$
 (d) $\{\ \}, \{a\}, \{b\}, \{a,b\}$
 (e) $\{\ \}, \{5\}, \{7\}, \{5,7\}$
 (f) $\{\ \}, \{5\}, \{e\}\{5,e\}$

 (g) $\{\ \}, \{2\}, \{4\}, \{6\}, \{2,4\},$
     $\{2,6\}, \{4,6\}, \{2,4,6\}$
 (h) $\{\ \}, \{x\}, \{y\}, \{z\}, \{x,y\},$
     $\{x,z\}, \{y,z\}, \{x,y,z\}$
 (i) $\{\ \}, \{2\}, \{b\}, \{c\}, \{2,b\},$
     $\{2,c\}, \{b,c\}, \{2,b,c\}$

Question 4.5:
 (a) $\Big\{\ \Big\}, \Big\{\{a,b,c\}\Big\}$
 (b) $\Big\{\ \Big\}, \Big\{\{x,y\}\Big\}$
 (c) $\Big\{\ \Big\}, \Big\{\{1,2,3\}\Big\}$
 (d) $\Big\{\ \Big\}, \Big\{r\Big\}, \Big\{\{s,t\}\Big\}, \Big\{r,\{s,t\}\Big\}$
 (e) $\Big\{\ \Big\}, \Big\{\{x\}\Big\}, \Big\{\{y,z\}\Big\}, \Big\{\{x\},\{y,z\}\Big\}$
 (f) $\Big\{\ \Big\}, \Big\{\{2,4\}\Big\}, \Big\{6\Big\}, \Big\{\{2,4\},6\Big\}$

 (g) $\Big\{\ \Big\}, \Big\{2\Big\}, \Big\{\{4\}\Big\}, \Big\{6\Big\}, \Big\{2,\{4\}\Big\},$
     $\Big\{2,6\Big\}, \Big\{\{4\},6\Big\}, \Big\{2,\{4\},6\Big\}$
 (h) $\Big\{\ \Big\}, \Big\{\{2\}\Big\}, \Big\{\{4\}\Big\}, \Big\{\{6\}\Big\}, \Big\{\{2\},\{4\}\Big\},$
     $\Big\{\{2\},\{6\}\Big\}, \Big\{\{4\},\{6\}\Big\}, \Big\{\{2\},\{4\},\{6\}\Big\}$
 (i) $\Big\{\ \Big\}, \Big\{u\Big\}, \Big\{v\Big\}, \Big\{\{w\}\Big\}, \Big\{u,v\Big\},$
     $\Big\{u,\{w\}\Big\}, \Big\{v,\{w\}\Big\}, \Big\{u,v,\{w\}\Big\}$

Question 4.6:
 (a) $\{2,3,4,5,6,7,8\}$
 (b) $\{4,5,6\}$
 (c) $\{0,1,7,8,9\}$
 (d) $\{0,1,2,3,9\}$
 (e) $\{0,1,9\}$
 (f) $\{0,1,2,3,7,8,9\}$
 (g) $\{0,1,2,3,7,8,9\}$
 (h) $\{0,1,9\}$
 (i) $\{2,3\}$
 (j) $\{7,8\}$

Question 4.7:
 (a) $\{1,2,3,4,a,b,c,d,e\}$
 (b) $\{2,c,d\}$
 (c) $\{3,4,5,e\}$
 (d) $\{1,5,a,b\}$
 (e) $\{5\}$
 (f) $\{1,3,4,5,a,b,e\}$
 (g) $\{1,3,4,5,a,b,e\}$
 (h) $\{5\}$
 (i) $\{1,a,b\}$
 (j) $\{3,4,e\}$

Question 4.8:
 (a) $\{o,p,r,s,t,u,v,y\}$
 (b) $\{p,s,t,u\}$
 (c) $\{q,r,v,w,x,z\}$
 (d) $\{o,q,w,x,y,z\}$
 (e) $\{q,w,x,z\}$
 (f) $\{o,q,r,v,w,x,y,z\}$
 (g) $\{o,q,r,v,w,x,y,z\}$
 (h) $\{q,w,x,z\}$
 (i) $\{o,y\}$
 (j) $\{r,v\}$

Question 4.9:
 (a) $\{-2,2\}$
 (b) $\{-1,1\}$
 (c) $\{-4,5\}$
 (d) $\{3\}$
 (e) $\{-5,5\}$
 (f) $\{-3,4\}$
 (g) $\{-2,2\}$
 (h) $\{-3,2\}$
 (i) $\{0\}$

Question 4.10:
 (a) $\{-1\}$
 (b) $\left\{\frac{-7}{4}, \frac{3}{2}\right\}$
 (c) $\left\{\frac{-5}{2}, \frac{-4}{3}\right\}$
 (d) $\{-2,2,3\}$
 (e) $\left\{\frac{-7}{3}, 5\right\}$
 (f) $\{1\}$
 (g) $\{0\}$
 (h) $\left\{\frac{4}{3}, \frac{2}{3}\right\}$
 (i) $\{-5,-3,3\}$

Question 4.11:

(a) $|A^c| = 100$

(b) $|B^c| = \infty$

(c) $|A^c \cap B^c| = 0$

(d) $|O^c| = \infty$

(e) $|O \cap E| = 0$

(f) $|E^c| = \infty$

(g) $|A \cup B| = \infty$

(h) $|B \cap E| = 150$

(i) $|A^c \cap O| = 50$

(j) $|O \cap E^c| = \infty$

(k) $|A \cap O^c| = \infty$

(l) $|B \cup E| = \infty$

Question 4.12: subsets: $2^6 = 64$, proper subsets: $2^6 - 1 = 63$

Question 4.13: subsets: $2^8 = 256$, proper subsets: $2^8 - 1 = 255$

Question 4.14: $2^{10} = 1,024$

Question 4.15: $2^{20} = 1,048,576$

Question 4.16: $A \times B = \{(a,1),(a,2),(a,3),(b,1),(b,2),(b,3),(c,1),(c,2),(c,3)\}$

Question 4.17: $S \times T = \{(e,9),(e,8),(e,7),(e,6),(f,9),(f,8),(f,7),(f,6),$
$(g,9),(g,8),(g,7),(g,6),(h,9),(h,8),(h,7),(h,6)\}$

Question 4.18: $X \times Y = \{(0,t),(1,t),(2,t),(3,t),(4,t),(5,t),$
$(0,f),(1,f),(2,f),(3,f),(4,f),(5,f)\}$

Question 4.19: $X_1 \times X_2 \times X_3 \times X_4 = \{(0,0,0,0),(0,0,0,1),(0,0,1,0),(0,0,1,1),$
$(0,1,0,0),(0,1,0,1),(0,1,1,0),(0,1,1,1),$
$(1,0,0,0),(1,0,0,1),(1,0,1,0),(1,0,1,1),$
$(1,1,0,0),(1,1,0,1),(1,1,1,0),(1,1,1,1)\}$

Question 4.20:

$A = \{e,m,a,x,j,c,w\}$

$B = \{j,c,w,k,i,b,v\}$

$A \cup B = \{e,m,a,x,j,c,w,k,i,b,v\}$

$A \cap B = \{j,c,w\}$

$A^c = \{g,t,d,o,k,i,b,v\}$

$B^c = \{g,t,d,o,e,m,a,x\}$

$(A \cup B)^c = \{g,t,d,o\}$

$(A \cap B)^c = \{g,t,d,o,e,m,a,x,k,i,b,v\}$

$A^c \cup B = \{g,t,d,o,j,c,w,k,i,v,b\}$

$A \cup B^c = \{g,t,d,o,e,m,a,x,j,c,w\}$

$A - B = \{e,m,a,x\}$

$B - A = \{k,i,b,v\}$

Question 4.21:

$A = \{g,i,d,t,b\}$

$B = \{t,b,p,w,v\}$

$A \cup B = \{g,i,d,t,b,p,w,v\}$

$A \cap B = \{t,b\}$

$A^c = \{m,j,z,c,l,p,w,v\}$

$B^c = \{m,j,z,c,l,g,i,d\}$

$(A \cup B)^c = \{m,j,z,c,l\}$

$(A \cap B)^c = \{m,j,z,c,l,g,i,d,p,w,v\}$

$A^c \cup B = \{m,j,z,c,l,t,b,p,w,v\}$

$A \cup B^c = \{m,j,z,c,l,g,i,d,t,b\}$

$A - B = \{g,i,d\}$

$B - A = \{p,w,v\}$

Question 4.22:

$A = \{c,z,p,r,g,n,t\}$

$B = \{h,d,q,p,r,s,t\}$

$C = \{v,o,a,g,n,s,t\}$

$A \cap B = \{p,r,t\}$

$A \cap C = \{g, n, t\}$
$B \cap C = \{s, t\}$
$A \cup B = \{c, z, p, r, g, n, t, s, h, q, d\}$
$A \cup C = \{c, z, p, r, g, n, t, s, v, o, a\}$
$B \cup C = \{h, q, d, p, r, s, t, g, n, v, o, a\}$
$A \cap B \cap C = \{t\}$
$A \cup B \cup C = \{c, z, p, r, g, n, t, h, d, q, s, v, o, a\}$
$A^c = \{x, f, y, k, j, v, o, a, s, h, q, d\}$
$B^c = \{x, f, y, k, j, c, z, g, n, v, o, a\}$
$C^c = \{x, f, y, k, j, c, z, p, r, h, d, q\}$
$(A \cap B)^c = \{x, f, y, k, j, c, z, g, n, v, o, a, s, h, q, d\}$
$(A \cap C)^c = \{x, f, y, k, j, c, z, p, r, h, q, d, s, v, o, a\}$
$(B \cap C)^c = \{x, f, y, k, j, d, q, h, r, p, c, z, g, n, v, o, a\}$
$(A \cup B)^c = \{x, f, y, k, j, v, o, a\}$
$(A \cup C)^c = \{x, f, y, k, j, h, q, d\}$
$(B \cup C)^c = \{x, f, y, k, j, c, z\}$
$(A \cap B \cap C)^c = \{x, f, y, k, j, c, z, p, r, g, n, h, q, d, s, v, o, a\}$
$(A \cup B \cup C)^c = \{x, f, y, k, j\}$
$A - B = \{c, z, g, n\}$
$B - A = \{h, d, q, s\}$
$A - C = \{c, z, p, r\}$
$C - A = \{v, o, a, s\}$
$B - C = \{p, r, h, q, d\}$
$C - B = \{g, n, v, o, a\}$

Question 4.23:
$A = \{i, p, j, m, t, c, y, n, g, l\}$
$B = \{w, k, s, o, t, c, y, d, l\}$
$C = \{q, a, v, u, gn, g, d, l\}$
$A \cap B = \{t, c, y, l\}$
$A \cap C = \{n, g, l\}$
$B \cap C = \{d, l\}$
$A \cup B = \{p, i, j, m, t, c, y, n, g, l, k, w, s, o, d\}$
$A \cup C = \{p, i, j, m, t, c, y, n, g, l, d, v, a, q, u\}$
$B \cup C = \{w, k, s, o, t, c, y, d, l, n, g, v, a, q, u\}$
$A \cap B \cap C = \{l\}$
$A \cup B \cup C = \{p, i, j, m, t, c, y, n, g, l, k, w, s, o, d, v, a, q, u\}$
$A^c = \{e, x, f, r, h, b, z, w, k, s, o, d, u, a, q, u\}$
$B^c = \{e, x, f, r, h, b, z, p, i, j, m, n, g, v, a, q, u\}$
$C^c = \{e, x, f, r, h, b, z, p, i, j, m, t, c, y, k, w, s, o\}$
$(A \cap B)^c = \{e, x, f, r, h, b, z, p, i, j, mn, g, v, a, q, u, d, w, k, s, o\}$
$(A \cap C)^c = \{e, x, f, r, h, b, z, p, i, j, m, t, c, y, w, k, s, o, d, v, a, q, u\}$
$(B \cap C)^c = \{e, x, f, r, h, b, z, p, i, j, m, t, c, y, k, w, s, o, n, g, v, a, q, u\}$
$(A \cup B)^c = \{e, x, f, r, h, b, z, v, a, q, u\}$
$(A \cup C)^c = \{e, x, f, r, h, b, z, w, k, s, o\}$
$(B \cup C)^c = \{e, x, f, r, h, b, z, p, i, m, j\}$
$(A \cap B \cap C)^c = \{e, x, f, r, h, b, z, p, i, j, m, t, c, y, k, w, s, o, n, g, v, a, q, u, d\}$
$(A \cup B \cup C)^c = \{e, x, f, r, h, b, z\}$
$A - B = \{i, j, p, mn, g\}$
$B - A = \{w, k, s, o, d\}$
$A - C = \{i, j, p, m, t, c, y\}$

$C - A = \{a, v, q, u, d\}$
$B - C = \{t, c, y, k, w, s, o\}$
$C - B = \{n, g, a, v, q, u\}$

Question 4.24:

$|A| = 27$	$|A \cup B \cup C| = 57$	$|(A \cap B \cap C)^c| = 69$
$|B| = 40$	$|A^c| = 47$	$|(A \cup B \cup C)^c| = 17$
$|C| = 26$	$|B^c| = 34$	$|A - B| = 13$
$|A \cap B| = 14$	$|C^c| = 48$	$|B - A| = 26$
$|A \cap C| = 11$	$|(A \cap B)^c| = 60$	$|A - C| = 16$
$|B \cap C| = 16$	$|(A \cap C)^c| = 63$	$|C - A| = 15$
$|A \cup B| = 53$	$|(B \cap C)^c| = 58$	$|B - C| = 24$
$|A \cup C| = 42$	$|(A \cup B)^c| = 21$	$|C - B| = 10$
$|B \cup C| = 50$	$|(A \cup C)^c| = 32$	
$|A \cap B \cap C| = 5$	$|(B \cup C)^c| = 24$	

Question 4.25:

(a) $A \cup A = A$

(b) $A \cup A^c = \mathcal{U}$

(c) $(A \cup A^c)^c = \varnothing$

(d) $A \cup (A \cap B) = A$

(e) $A \cup (A^c \cap B) = A \cup B$

(f) $A \cap (A^c \cup B) = A \cap B$

(g) $A \cap A = A$

(h) $A \cap (A \cup B) = A$

(i) $A \cap (A \cup B \cup C) = A$

(j) $(A \cap B) \cup (A \cap B^c) = A$

(k) $(A^c \cup A^c)^c = A$

(l) $(A \cap B) \cup (A^c \cap B) = B$

Question 4.26:

(a) $(A^c \cup B^c) \cap (A^c \cup B) = A^c$

(b) $B \cup (B \cap B^c) = B$

(c) $A^c \cup (B \cap A^c) = A^c$

(d) $(A \cup B^c) \cap (A \cup B) = A$

(e) $C \cup [C \cup (C \cap A)] = C$

(f) $A \cap [A \cup (A \cap B)] = A$

(g) $C \cup (C \cap A^c \cap B \cap D) = C$

(h) $C^c \cap (C \cap A \cap B \cap D)^c = C^c$

Question 4.27:

(a) $B \cap [A \cup (A^c \cap B)] = B$

(b) $[(A \cap B^c) \cup A^c]^c = A \cap B$

(c) $(A^c \cap B) \cup [A \cap (A \cup B)] = A \cup B$

(d) $[A \cup (A \cap B)]^c \cap B = A^c \cap B$

(e) $(A \cap C^c) \cup (A^c \cap B) \cup (B \cap C)^c = A^c \cup B^c \cup C^c$

(f) $(A \cap B)^c \cup (B \cap C) = A^c \cup B^c \cup C$

(g) $[A \cup (B \cap C)] \cap (A^c \cup C) = (A \cup B) \cap C$

(h) $(A \cap C^c) \cup (A^c \cap B) \cup (A \cap C)^c = A^c \cup C^c$

(i) $(A \cap C) \cup (A^c \cap B) \cup (C \cap B) = (A \cap C) \cup (B \cap A^c)$

(j) $A^c \cup B^c \cup [A \cap B \cap C^c] = A^c \cup B^c \cup C^c$

(k) $(A \cup C) \cap (A^c \cup B) \cap (C \cup B) = (A \cap B) \cup (C \cap A^c)$

(l) $(A \cap B)^c \cap (A^c \cup B) \cap (B^c \cup B) = A^c$

Question 4.28:

(a) $R = \{(1,1), (2,2), (3,3), (4,4)\}$

(b) $R = \{(1,2), (1,3), (1,4), (2,3), (2,4), (3,4)\}$

(c) $R = \{(2,1), (3,1), (3,2), (4,1), (4,2), (4,3)\}$

(d) $R = \{(1,1), (1,2), (1,3), (1,4), (2,2), (2,3), (2,4), (3,3), (3,4), (4,4)\}$

(e) $R = \{(1,1), (2,1), (2,2), (3,1), (3,2), (3,3), (4,1), (4,2), (4,3), (4,4)\}$

(f) $R = \{(1,2), (1,3), (1,4), (2,1), (2,3), (2,4), (3,1), (3,2), (3,4), (4,1), (4,2), (4,3)\}$

Question 4.29:
(a) $R = \{(2,2),(5,5)\}$
(b) $R = \{(0,1),(0,2),(0,4),(0,5),(0,6),(2,4),(2,5),(2,6),(3,4),(3,5),(3,6),(5,6)\}$
(c) $R = \{(2,1),(3,1),(3,2),(5,1),(5,2),(5,4),(7,1),(7,2),(7,4),(7,5),(7,6)\}$
(d) $R = \{(0,1),(0,2),(0,4),(0,5),(0,6),(2,2),(2,4),(2,5),(2,6),(3,4),(3,5),(3,6),(5,5),(5,6)\}$
(e) $R = \{(2,1),(2,2),(3,1),(3,2),(5,1),(5,2),(5,4),(5,5)(7,1),(7,2),(7,4),(7,5),(7,6)\}$
(f) $R = \{(0,1),(0,2),(0,4),(0,5),(0,6),(2,1),(2,4),(2,5),(2,6),$
$(3,1),(3,2),(3,4),(3,5),(3,6),(5,1),(5,2),(5,4),(5,6),(7,1),(7,2),(7,4),(7,5),(7,6)\}$

Question 4.30: Domain $= \{1,3\}$, Range $= \{2,3,4\}$, $R = \{(1,2),(1,3),(1,4),(3,3),(3,4)\} \subset$ Domain $\times$ Range

Question 4.31:
(a) reflexive, antisymmetric, transitive
(b) transitive
(c) reflexive, symmetric, antisymmetric, transitive
(d) irreflexive, symmetric
(e) transitive
(f) reflexive, antisymmetric, transitive

Question 4.32: Similar to Example 4.36.

Question 4.33: Similar to Example 4.37. Equivalence classes are $\{1,5\}, \{2,4,6\}, \{3\}$.

Question 4.34: This relation is reflexive, symmetric, antisymmetric, and transitive. Thus it is both a partial ordering and an equivalence relation.

B.5   CHAPTER FIVE ANSWERS

Question 5.1:
(a) $x + x = x$
(b) $x + x' = 1$
(c) $(x + x')' = 0$
(d) $x + xy = x$
(e) $x + x'y = x + y$
(f) $x(x' + y) = xy$
(g) $x(x) = x$
(h) $x(x + y) = x$
(i) $x(x + y + z) = x$
(j) $xy + xy' = x$
(k) $(x' + x')' = x$
(l) $xy + x'y = y$

Question 5.2:
(a) $(x' + y')(x' + y) = x'$
(b) $y + (yy') = y$
(c) $x' + yx' = x'$
(d) $(x + y')(x + y) = x$
(e) $w + [w + (wx)] = w$
(f) $x[x + (xy)] = x$
(g) $w + (wx'yz) = w$
(h) $w'(wxyz)' = w'$

Question 5.3:
(a) $y[x + (x'y)] = y$
(b) $[(xy') + x']' = xy$
(c) $x'y + x(x + y) = x + y$
(d) $(x + xy)'y = x'y$
(e) $xz' + x'y + (yz)' = x' + y' + z'$
(f) $(xy)' + (yz) = x' + y' + z$

(g) $[x + (yz)](x' + z) = (x + y)z$
(h) $xz' + x'y + (xz)' = x' + z'$
(i) $xz + x'y + zy = xz + x'y$
(j) $x' + y' + xyz' = x' + y' + z'$
(k) $(x + z)(x' + y)(z + y) = xy + x'z$
(l) $(xy)'(x' + y)(y' + y) = x'$

Question 5.4:
(a) $x'y$
(b) $x' + y$

(c) $(xy) + z'$ or $xy + z'$
(d) $(x + y)z'$

(e) $(x + y) + z'$ or $x + y + z'$
(f) $(xy)z'$ or xyz'

Question 5.5:
(a) $(xy') + y$ or $xy' + y$
(b) $(x + y')y$

(c) $(x + y') + y$ or $x + y' + y$
(d) $(xy')y$ or $xy'y$

Question 5.6:
(a) $[(x + y)z]'$
(b) $[(xy) + z]'$ or $(xy + z)'$

(c) $[(xy)z]'$ or $(xyz)'$
(d) $[(x + y) + z]$; or $(x + y + z)'$

Question 5.7:
(a) $(z'x) + y$ or $z'x + y$
(b) $(z' + x)y$

(c) $(z'x)y$ or $z'xy$
(d) $(z' + x) + y$ or $z' + x + y$

Question 5.8: Diagrams will vary.

Question 5.9: Diagrams will vary.

Question 5.10: Diagrams will vary.

Question 5.11:
(a) $x'y' + xy'$ and $(x + y')(x' + y')$
(b) $x'y' + x'y$ and $(x' + y)(x' + y')$
(c) $x'y$ and $(x + y)(x' + y)(x' + y')$
(d) $xy' + xy$ and $(x + y)(x + y')$
(e) $x'y' + xy$ and $(x + y')(x' + y)$

(f) xy and $(x + y)(x + y')(x' + y)$
(g) $x'y' + x'y + xy'$ and $x' + y'$
(h) $x'y' + xy' + xy$ and $x + y'$
(i) $x'y + xy$ and $(x + y)(x' + y)$

Question 5.12:
(a) $x'yz' + xy'z + xyz$ and
$(x + y + z)(x + y + z')(x + y' + z')(x' + y + z)(x' + y' + z)$
(b) $x'y'z' + x'yz' + x'yz + xy'z' + xyz'$ and
$(x + y + z')(x' + y + z')(x' + y' + z')$
(c) $x'y'z' + x'yz' + xy'z + xyz'$ and
$(x + y + z')(x + y' + z')(x' + y + z)(x' + y' + z')$
(d) $x'y'z' + x'yz + xyz'$ and
$(x + y + z')(x + y' + z)(x' + y + z)(x' + y + z')(x' + y' + z')$
(e) $x'y'z' + x'y'z + xy'z + xyz' + xyz$ and
$(x + y' + z)(x + y' + z')(x' + y + z)$
(f) $x'y'z' + x'y'z + x'yz + xyx$ and
$(x + y' + z)(x' + y + z)(x' + y + z')(x' + y' + z)$

Question 5.13:

(a)

x	y	$(x'+y)x$
0	0	0
0	1	0
1	0	0
1	1	1

(b)

x	y	$(x+y')x$
0	0	0
0	1	0
1	0	1
1	1	1

(c)

x	y	$(x+y)y'$
0	0	0
0	1	0
1	0	1
1	1	0

(d)

x	y	$(x+y)(x+y)$
0	0	0
0	1	1
1	0	1
1	1	1

(e)

x	y	$(x+y)(x'+y)$
0	0	0
0	1	1
1	0	0
1	1	1

(f)

x	y	$(x+y')(x'+y)$
0	0	1
0	1	0
1	0	0
1	1	1

(g)

x	y	$xx+x(x+y)$
0	0	0
0	1	0
1	0	1
1	1	1

(h)

x	y	$xy+y(x+y)$
0	0	0
0	1	1
1	0	0
1	1	1

(i)

x	y	$x'y+y'(x+y)$
0	0	0
0	1	1
1	0	1
1	1	0

(j)

x	y	$x'y'+y'(x+y)$
0	0	1
0	1	0
1	0	1
1	1	0

(k)

x	y	$xy(x+y)$
0	0	0
0	1	0
1	0	0
1	1	1

(l)

x	y	$x'y'(x'+y')$
0	0	1
0	1	0
1	0	0
1	1	0

Question 5.14:

(a)

x	y	z	$x(x+y'+z)$
0	0	0	0
0	0	1	0
0	1	0	0
0	1	1	0
1	0	0	1
1	0	1	1
1	1	0	1
1	1	1	1

(b)

x	y	z	$x'(x+y'+z)$
0	0	0	1
0	0	1	1
0	1	0	0
0	1	1	1
1	0	0	0
1	0	1	0
1	1	0	0
1	1	1	0

(c)

x	y	z	$(x+y)(x+z)$
0	0	0	0
0	0	1	0
0	1	0	0
0	1	1	1
1	0	0	1
1	0	1	1
1	1	0	1
1	1	1	1

(d)

x	y	z	$(x'+y')(x+z)$
0	0	0	0
0	0	1	1
0	1	0	0
0	1	1	1
1	0	0	1
1	0	1	1
1	1	0	0
1	1	1	0

(e)

x	y	z	$x'y' + z'(x+y)$
0	0	0	1
0	0	1	1
0	1	0	1
0	1	1	0
1	0	0	1
1	0	1	0
1	1	0	1
1	1	1	0

(f)

x	y	z	$xy(x+y+z)$
0	0	0	0
0	0	1	0
0	1	0	0
0	1	1	0
1	0	0	0
1	0	1	0
1	1	0	1
1	1	1	1

B.6   CHAPTER SIX ANSWERS

Question 6.1:
For function in Fig. 6.3(a):
  (a) Domain: $X = \{1,2,3,4,5\}$, Codomain: $Y = \{a,b,c,d,e\}$, Range: $R = \{a,b,c,d,e\}$
  (b) Both one-to-one and onto
  (c) $f(1)=c, f(2)=b, f(3)=a, f(4)=e, f(5)=d$
  (d) $\{(1,c),(2,b),(3,a),(4,e),(5,d)\} \subset X \times Y$

For function in Fig. 6.3(b):
  (a) Domain: $X = \{1,2,3,4,5\}$, Codomain: $Y = \{a,b,c,d,e\}$, Range: $R = \{b,d\}$
  (b) Neither one-to-one nor onto
  (c) $f(1)=b, f(2)=d, f(3)=d, f(4)=b, f(5)=d$
  (d) $\{(1,b),(2,d),(3,d),(4,b),(5,d)\} \subset X \times Y$

For function in Fig. 6.3(c):
  (a) Domain: $X = \{1,2,3,4,5\}$, Codomain: $Y = \{a,b,c,d,e\}$, Range: $R = \{a,b,c,d,e\}$
  (b) Both one-to-one and onto
  (c) $f(1)=d, f(2)=e, f(3)=a, f(4)=b, f(5)=c$
  (d) $\{(1,d),(2,e),(3,a),(4,b),(5,c)\} \subset X \times Y$

For function in Fig. 6.3(d):
  (a) Domain:$X = \{1,2,3,4,5,6,7\}$, Codomain:$Y = \{a,b,c,d,e,f,g\}$, Range:$R = \{c,d,f,g\}$
  (b) Neither one-to-one nor onto
  (c) $f(1)=c, f(2)=d, f(3)=g, f(4)=f, f(5)=c, f(6)=d, f(7)=f$
  (d) $\{(1,c),(2,d),(3,g),(4,f),(5,c),(6,d),(7,f)\} \subset X \times Y$

For function in Fig. 6.3(e):
  (a) Domain: $X = \{1,2,3,4,5,6,7\}$, Codomain: $Y = \{a,b,c,d,e,f,g\}$, Range: $R = \{a,b,c,d,f,g\}$
  (b) Neither one-to-one nor onto
  (c) $f(1)=c, f(2)=d, f(3)=a, f(4)=g, f(5)=b, f(6)=c, f(7)=f$
  (d) $\{(1,c),(2,d),(3,a),(4,g),(5,b),(6,c),(7,f)\} \subset X \times Y$

For function in Fig. 6.3(f):
  (a) Domain: $X = \{1,2,3,4,5,6,7\}$, Codomain: $Y = \{a,b,c,d,e,f,g\}$, Range: $R = \{a,b,c,d,e,f,g\}$
  (b) Both one-to-one and onto
  (c) $f(1)=e, f(2)=b, f(3)=a, f(4)=c, f(5)=d, f(6)=g, f(7)=f$
  (d) $\{(1,e),(2,b),(3,a),(4,c),(5,d),(6,g),(7,f)\} \subset X \times Y$

For function in Fig. 6.3(g):
  (a) Domain:$X = \{1, 2, 3, 4, 5, 6, 7\}$, Codomain:$Y = \{a, b, c, d, e, f, g\}$, Range:$R = \{a, c, d, f\}$
  (b) Neither one-to-one nor onto
  (c) $f(1) = d, f(2) = d, f(3) = f, f(4) = f, f(5) = c, f(6) = c, f(7) = a$
  (d) $\{(1, d), (2, d), (3, f), (4, f), (5, c), (6, c), (7, a)\} \subset X \times Y$

For function in Fig. 6.3(h):
  (a) Domain: $X = \{1, 2, 3, 4, 5, 6, 7\}$, Codomain: $Y = \{a, b, c, d, e, f, g\}$, Range: $R = \{a, b, c, d, e, f, g\}$
  (b) Both one-to-one and onto
  (c) $f(1) = b, f(2) = c, f(3) = d, f(4) = a, f(5) = f, f(6) = g, f(7) = e$
  (d) $\{(1, b), (2, c), (3, d), (4, a), (5, f), (6, g), (7, e)\} \subset X \times Y$

For function in Fig. 6.3(i):
  (a) Domain:$X = \{1, 2, 3, 4, 5, 6, 7\}$, Codomain:$Y = \{a, b, c, d, e, f, g\}$, Range:$R = \{a, b, c, f\}$
  (b) Neither one-to-one nor onto
  (c) $f(1) = b, \ f(2) = a, \ f(3) = c, \ f(4) = b, \ f(5) = a, \ f(6) = f, \ f(7) = f$
  (d) $\{(1, b), (2, a), (3, c), (4, b), (5, a), (6, f), (7, f)\} \subset X \times Y$

Question 6.2:
For function f_1:
  (a) Range: $R = \{b, c, e\}$
  (b) Neither one-to-one nor onto

  (c)
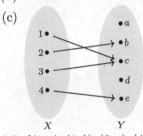
  (d) $\{(1, c), (2, b), (3, c), (4, e)\} \subset X \times Y$

For function f_3:
  (a) Range: $R = \{b, d\}$
  (b) Neither one-to-one nor onto

  (c)
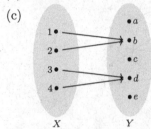
  (d) $\{(1, b), (2, b), (3, d), (4, d)\} \subset X \times Y$

For function f_2:
  (a) Range: $R = \{b, c, d, e\}$
  (b) One-to-one but not onto onto

  (c)
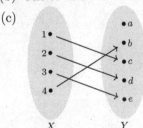
  (d) $\{(1, c), (2, d), (3, e), (4, b)\} \subset X \times Y$

Question 6.3:

For function f_1:

  (a) Range: $R = \{a, b, c, d, e, f\}$

  (b) Onto but not one-to-one

  (c)

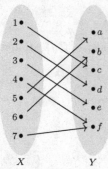

  (d) $\{(1, c), (2, d), (3, e), (4, f), (5, a), (6, b), (7, f)\} \subset X \times Y$

For function f_2:

  (a) Range: $R = \{a, b, c, d, e, f\}$

  (b) Onto but not one-to-one

  (c)

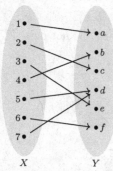

  (d) $\{(1, a), (2, c), (3, e), (4, b), (5, d), (6, f), (7, d)\} \subset X \times Y$

For function f_3:

  (a) Range: $R = \{b, c, d, e\}$

  (b) Neither onto nor one-to-one

  (c)

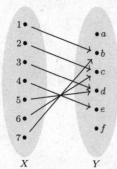

  (d) $\{(1, b), (2, c), (3, d), (4, e), (5, d), (6, c), (7, b)\} \subset X \times Y$

Question 6.4:
For function f_1:
  (a) Range: $R = \{a, b, c, d\}$
  (b) Neither one-to-one nor onto

  (c)
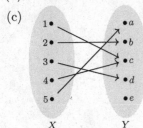

  X Y
  (d) $f(1) = c, f(2) = b, f(3) = d, f(4) = c, f(5) = a$

For function f_2:
  (a) Range: $R = \{a, b, c, d\}$
  (b) Neither one-to-one nor onto

  (c)
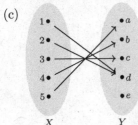

  X Y
  (d) $f(1) = d, f(2) = d, f(3) = c, f(4) = b, f(5) = a$

For function f_3:
  (a) Range: $R = \{a, b, c, d, e\}$
  (b) One-to-one and onto

  (c)
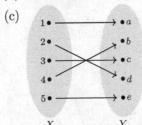

  X Y
  (d) $f(1) = a, f(2) = d, f(3) = c, f(4) = b, f(5) = e$

Question 6.5:

For function f_1:

(a) Range: $R = \{a, b, c, d, e, f\}$

(b) One-to-one and onto

(c)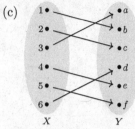

(d) $f(1)=b, f(2)=c, f(3)=a, f(4)=e, f(5)=f, f(6)=d$

For function f_2:

(a) Range: $R = \{b, e\}$

(b) Neither one-to-one nor onto

(c)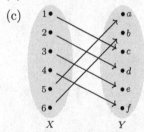

(d) $f(1)=b, f(2)=b, f(3)=b, f(4)=e, f(5)=e, f(6)=e$

For function f_3:

(a) Range: $R = \{a, b, c, d, e, f\}$

(b) One-to-one and onto

(c)

(d) $f(1)=c, f(2)=d, f(3)=e, f(4)=f, f(5)=a, f(6)=b$

Question 6.6:

For function in Fig. 6.4(a):
(a) Domain: $X = \{-3, -2, 0, 2, 4\}$,
   Range: $Y = \{-3, -2, 1, 2, 4\}$

(b)

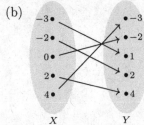

X Y

(c) $f(-3)=1, f(-2)=2, f(0)=-2, f(2)=4, f(4)=-3$
(d) $\{(-3,1),(-2,2),(0,-2),(2,4),(4,-3)\} \subset X \times Y$

For function in Fig. 6.4(b):
(a) Domain: $X = \{-4, -3, -2, 0, 2, 3, 4\}$,
   Range: $Y = \{-4, -3, -2, 0, 2, 3, 4\}$

(b)

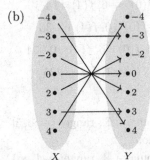

X Y

(c) $f(-4)=4, f(-3)=-3, f(-2)=2, f(0)=0, f(2)=-2, f(3)=3, f(4)=-4$
(d) $\{(-4,4),(-3,-3),(-2,2),(0,0),(2,-2),(3,3),(4,-4)\} \subset X \times Y$

For function in Fig. 6.4(c):
(a) Domain: $X = \{-4, -2, 0, 1, 2, 4\}$,
   Range: $Y = \{-4, -3, 2, 3, 4, 5\}$

(b)

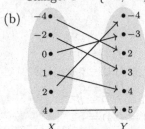

X Y

(c) $f(-4)=2, f(-2)=3, f(0)=-3, f(1)=4, f(2)=-4, f(4)=5$
(d) $\{(-4,2),(-2,3),(0,-3),(1,4),(2,-4),(4,5)\} \subset X \times Y$

For function in Fig. 6.4(d):
  (a) Domain: $X = \{-3, -2, -1, 0, 1, 2, 3, 4, \}$,
      Range: $Y = \{-4, -3, -2, -1, 0, 1, 2, 3, \}$
  (b)

  (c) $f(-3)=-4, f(-2)=-3, f(-1)=-2, f(0)=-1, f(1)=0, f(2)=1, f(3)=2, f(4)=3$
  (d) $\{(-3, -4), (-2, -3), (-1, -2), (0, -1), (1, 0), (2, 1), (3, 2), (4, 3)\} \subset X \times Y$

Question 6.7:

  (a) $f(-3)=-2, f(0)=7, f(4)=19$
  (b) $f(-3)=-2, f(0)=7, f(4)=19$
  (c) $f(-3)=18, f(0)=6, f(4)=-10$
  (d) $f(-3)=0.125, f(0)=1, f(4)=16$
  (e) $f(-3)=0, f(0)=0, f(4)=28$

  (f) $f(-3)=0, f(0)=-6, f(4)=14$
  (g) $f(-3)=48, f(0)=0, f(4)=-36$
  (h) $f(-3)=162, f(0)=6, f(4) \approx 0.074$
  (i) $f(-3)=64, f(0)=1, f(4)=225$

Question 6.8:

  (a) not a function
  (b) function, domain: $\mathbb{R}$, range: $[-4, \infty)$
  (c) function, domain: $\mathbb{R}$, range: $\mathbb{R}$

  (d) function, domain: $\mathbb{R}$, range: $(-\infty, 5]$
  (e) not a function
  (f) function, domain: $\mathbb{R}$, range: $[-4, \infty)$

Question 6.9:
$g \circ f(1) = g(f(1)) = g(c) = \alpha$
$g \circ f(2) = g(f(2)) = g(b) = \gamma$
$g \circ f(3) = g(f(3)) = g(d) = \beta$
$g \circ f(4) = g(f(4)) = g(a) = \delta$

Question 6.10:
$g \circ f(1) = g(f(1)) = g(c) = 4$
$g \circ f(2) = g(f(2)) = g(b) = 4$
$g \circ f(3) = g(f(3)) = g(a) = 2$
$g \circ f(4) = g(f(4)) = g(e) = 4$
$g \circ f(5) = g(f(5)) = g(d) = 2$
Domain: $\{1, 2, 3, 4, 5\}$,
Codomain: $\{1, 2, 3, 4, 5\}$,
Range: $\{2, 4\}$

$f \circ g(a) = f(g(a)) = f(2) = b$
$f \circ g(b) = f(g(b)) = f(4) = e$
$f \circ g(c) = f(g(c)) = f(4) = e$
$f \circ g(d) = f(g(d)) = f(2) = b$
$f \circ g(e) = f(g(e)) = f(4) = e$
Domain: $\{a, b, c, d, e\}$,
Codomain: $\{a, b, c, d, e\}$,
Range: $\{b, e\}$

Question 6.11:
For function in Fig. 6.3(a):
$f^{-1}(a)=3, f^{-1}(b)=2, f^{-1}(c)=1, f^{-1}(d)=5, f^{-1}(e)=4$
For function in Fig. 6.3(b):
f^{-1} does not exist
For function in Fig. 6.3(c):
$f^{-1}(a)=3, f^{-1}(b)=4, f^{-1}(c)=5, f^{-1}(d)=1, f^{-1}(e)=1$
For function in Fig. 6.3(d):
f^{-1} does not exist
For function in Fig. 6.3(e):
f^{-1} does not exist
For function in Fig. 6.3(f):
$f^{-1}(a)=3, f^{-1}(b)=2, f^{-1}(c)=4, f^{-1}(d)=5, f^{-1}(e)=1, f^{-1}(f)=7, f^{-1}(g)=6$
For function in Fig. 6.3(g):
f^{-1} does not exist
For function in Fig. 6.3(h):
$f^{-1}(a)=4, f^{-1}(b)=1, f^{-1}(c)=2, f^{-1}(d)=3, f^{-1}(e)=7, f^{-1}(f)=5, f^{-1}(g)=6$
For function in Fig. 6.3(i):
f^{-1} does not exist

Question 6.12:
f_1^{-1} and f_3^{-1} do not exist; $f_2^{-1}(b)=4, f_2^{-1}(c)=1, f_2^{-1}(d)=2, f_2^{-1}(e)=3$

Question 6.13:
$f_1^{-1}, f_2^{-1},$ and f_3^{-1} do not exist

Question 6.14:
f_1^{-1} and f_2^{-1} do not exist; $f_3^{-1}(a)=1, f_3^{-1}(b)=4, f_3^{-1}(c)=3, f_3^{-1}(d)=2, f_3^{-1}(e)=5$

Question 6.15:
  (a) Yes   (b) No   (c) Yes   (d) No

Question 6.16:
  (a) $f^{-1}(x) = \frac{x}{7}$ (c) $f^{-1}(x) = \frac{2-x}{5}$ (e) $f^{-1}(x) = (x+4)^2$
  (b) $f^{-1}(x) = \frac{1}{x+3}$ (d) $f^{-1}(x) = \frac{3x-10}{8}$ (f) $f^{-1}(x) = \frac{2}{x} + 3$

B.7   CHAPTER SEVEN ANSWERS

Question 7.1: There are five odd decimal digits (1,3,5,7,9) so $5^5 = 3\,125$.

Question 7.2: There are four odd octal digits (1,3,5,7) so $4^5 = 1\,025$.

Question 7.3: There are eight odd hexadecimal digits (1,3,5,7,9,B,D,F) so $8^5 = 32\,768$.

Question 7.4: $P_3^{32} = \frac{32!}{29!} = 32 \times 31 \times 30 = 29\,760$

Question 7.5: $P_5^{20} = \frac{20!}{15!} = 20 \times 19 \times 18 \times 17 \times 16 = 1\,860\,480$

Question 7.6: $\binom{20}{5} = \frac{20!}{5!15!} = 15\,504$

Question 7.7: $(8+6) \times (5+9) = 196$

Question 7.8: $10^4 + 10^3 = 11\,000$

Question 7.9:

(a) 5 040   (b) 2 520   (c) 907 200

Question 7.10: Number of functions: 531 441; Number of one-to-one functions: 60 480

Question 7.11: Number of functions: 14 641; Number of one-to-one functions: 7 920

Question 7.12: $C_2^{10} \cdot C_2^{10} \cdot C_3^{20} \cdot C_3^{20} = 2\,631\,690\,000$

Question 7.13: **while-do** in step 4: $n - 1$

Question 7.14: **for-do** in step 2: n; **for-do** in step 2.1: n^2; **for-do** in step 2.1.2: n^3

Question 7.15: **for-do** in step 2: $n - 1$; **while-do** in step 2.3: $\binom{n}{2} = 0.5n^2 - 0.5n$

Question 7.16: **for-do** in step 2: n; **for-do** in step 3: n; **for-do** in step 3.1: n^2; **for-do** in step 3.1.1.1: n^3

B.8 CHAPTER EIGHT ANSWERS

Question 8.1:

1.1 Multiplication in step 2.

1.2 Comparison $>$ in step 2.

1.3 Comparison $<$ in step 2.

1.4 Addition and subtraction in steps 3.1 and 3.2.

1.5 Additions in steps 3.1 and 3.2.1.

1.6 Multiplication in step 3.1.

1.7 Multiplication in step 3.2.

1.8 Multiplication in step 3.1.

1.9 Addition (sum) in step 3.1.

1.10 Comparison $=$ in step 2.1.

1.11 Comparisons $=$ in steps 2.1 and 3.1.

1.12 Comparison $=$ in step 2.1.1.

1.13 Comparison $=$ in step 2.1.1.

1.14 Multiplication $i \times i$ in step 3.1.

1.15 Division in step 3.1.

1.16 Comparison $<$ in step 3.1.

1.17 Comparisons (is x_i a digit) in step 4.1.

Question 8.2: $f(n) = 7n \leq 7 \cdot g(n)$

Question 8.3: $f(n) = 25n + 500 \leq 25n + 500n = 525n = 525 \cdot g(n)$

Question 8.4: $f(n) = 20\sqrt{n} + 40n^2 + 500n \log_2(n) \leq 20n^2 + 40n^2 + 500n^2 = 560n^2$ so $g(n) = n^2$

Question 8.5: $f(n) = 800n^2 + 40n^3 + 50 \log_2(n) \leq 800n^3 + 40n^3 + 50n^3 = 890n^3$ so $g(n) = n^3$.

Question 8.6:

(a) Algorithm one is faster.

(b) Algorithm two is faster.

(c) Algorithms are equally fast.

(d) Algorithm one is faster.

(e) Algorithm two is faster.

(f) Algorithm one is faster.

(g) Algorithm one is faster.

(h) Algorithms are equally fast.

Question 8.7:

(a) Algorithm one is faster.

(b) Algorithm two is faster.

(c) Algorithm two is faster.

(d) Algorithm one is faster.

(e) The algorithms are equally fast.

(f) Algorithm two is faster.

(g) Algorithm two is faster.

Question 8.8: $\sqrt{n} < n^{1.5} < 10^n$

Question 8.9: $n^2 \log_2(n) < 2^n < 2^{2^n}$

Question 8.10: $\log_2(n) < n < n \log_2(n)$ so $2^{\log_2(n)} < 2^n < 2^{n \log_2(n)}$

Question 8.11: The dominant operation is the comparison in step 4.1 and the time complexity is $f(n) = n - 1$.

Question 8.12: The dominant operation is the multiplication in step 2.1.2.1 and the time complexity is $f(n) = n^3$.

Question 8.13: The dominant operation is the addition in step 3.1.1.1.1 and the time complexity is $f(n) = n^3$.

B.9   CHAPTER NINE ANSWERS

Question 9.1:

For graph in Fig. 9.9(a):

(a) $\mathcal{V} = \{A, B, C, D, E\}$
$\mathcal{E} = \{\overline{AB}, \overline{AE}, \overline{AE}, \overline{BD}, \overline{BD}, \overline{CE}, \overline{DE}\}$

(b) The graph's order is five.

(c) $\deg(A)=3$, $\deg(B)=3$, $\deg(C)=1$, $\deg(D)=3$, $\deg(E)=4$

(d) Vertices A and D are adjacent to vertex B.

(e) Edge $\overline{CE}$ is incident to vertex C.

(f) There are no loops.

(g) There are four parallel edges; $\overline{AE}$, $\overline{AE}$, $\overline{BD}$, $\overline{BD}$.

(h) There is one bridge; $\overline{CE}$.

For graph in Fig. 9.9(b):

(a) $\mathcal{V} = \{A, B, C, D\}$
$\mathcal{E} = \{\overline{AA}, \overline{AB}, \overline{BD}, \overline{BD}, \overline{BD}, \overline{CC}, \overline{CD}\}$

(b) The graph's order is four.

(c) $\deg(A)=3$, $\deg(B)=4$, $\deg(C)=3$, $\deg(D)=4$

(d) Vertices A and D are adjacent to vertex B.

(e) Edges $\overline{CC}$ and $\overline{CD}$ are incident to vertex C.

(f) There are two loops; $\overline{AA}$ and $\overline{CC}$.

(g) There are three parallel edges; $\overline{BD}$, $\overline{BD}$, and $\overline{BD}$.

(h) There are two bridges; $\overline{AB}$ and $\overline{CD}$.

For graph in Fig. 9.9(c):

(a) $\mathcal{V} = \{A, B, C, D, E\}$
$\mathcal{E} = \{\overline{AA}, \overline{AB}, \overline{BD}, \overline{CD}, \overline{CE}, \overline{CE}, \overline{CE}, \overline{DE}\}$

(b) The graph's order is five.

(c) $\deg(A) = 3$, $\deg(B) = 2$, $\deg(C) = 4$, $\deg(D) = 3$, $\deg(E) = 4$

(d) Vertices A and D are adjacent to vertex B.

(e) Edges $\overline{CD}$, $\overline{CE}$, $\overline{CE}$, $\overline{CE}$ are incident to vertex C.

(f) There is one loops; $\overline{AA}$.

(g) There are three parallel edges; $\overline{CE}$, $\overline{CE}$, and $\overline{CE}$.

(h) There are two bridges; $\overline{AB}$ and $\overline{BD}$.

For graph in Fig. 9.9(d):

(a) $\mathcal{V} = \{A, B, C, D, E, F\}$
$\mathcal{E} = \{\overline{AB}, \overline{AC}, \overline{BC}, \overline{DE}, \overline{DF}, \overline{EF}\}$

(b) The graph's order is six.

(c) $\deg(A) = 2$, $\deg(B) = 2$, $\deg(C) = 2$, $\deg(D) = 2$, $\deg(E) = 2$, $\deg(F) = 2$

(d) Vertices A and C are adjacent to vertex B.

(e) Edges $\overline{AC}$ and $\overline{BC}$ are incident to vertex C.

(f) There are no loops.

(g) There are no parallel edges.

(h) This graph is not connected and therefore cannot have a bridge.

Question 9.2:

(a) Not adjacent   (d) Adjacent   (g) Adjacent

(b) Not adjacent   (e) Adjacent   (h) Adjacent

(c) Adjacent   (f) Not adjacent   (i) Not adjacent

Question 9.3: The graph is not connected; it has two components.

Question 9.4: $\deg(A) = 4$, $\deg(B) = 2$, $\deg(C) = 3$, $\deg(D) = 5$, $\deg(E) = 2$; there are two loops, $\overline{AA}$ and $\overline{DD}$; there are no parallel edges; there are no bridges

Question 9.5: $\deg(A) = 4$, $\deg(B) = 3$, $\deg(C) = 2$, $\deg(D) = 4$, $\deg(E) = 2$, $\deg(F) = 3$; There are two loops, $\overline{AA}$ and $\overline{FF}$; there are two parallel edges, $\overline{BD}$ and $\overline{BD}$; there are three bridges, $\overline{AD}$, $\overline{AE}$, and $\overline{EF}$

Question 9.6: Three components

Question 9.7: Nine edges

Question 9.8: Eighteen edges

Question 9.9: Degree four

Question 9.10: Degree five

Question 9.11:
(a) Neither
(b) Semi-Eulerian
(c) Semi-Eulerian
(d) Neither

Question 9.12:
(a) Neither
(b) Semi-Eulerian
(c) Eulerian
(d) Semi-Eulerian

Question 9.13: Semi-Eulerian

Question 9.14: One possible path: $A - B - C - C - D - F - D - B - E - A - E - C$

Question 9.15: One possible path: $B - C - D - B - D - A - A - E - F - F$

Question 9.16: The trace of the modified Fleury's algorithm is given below:

Step	current_path	insertion_point	e	v	new_path	unused_edges
2	A	-	-	-	-	$\{e_1, e_2, e_3, e_4, e_5, e_6\}$
3.1	A	A	-	-	-	$\{e_1, e_2, e_3, e_4, e_5, e_6\}$
3.2	A	A	-	A	A	$\{e_1, e_2, e_3, e_4, e_5, e_6\}$
3.3.1–4	A	A	e_1	B	Ae_1B	$\{e_2, e_3, e_4, e_5, e_6\}$
3.3.1–4	A	A	e_3	C	Ae_1Be_3C	$\{e_2, e_4, e_5, e_6\}$
3.3.1–4	A	A	e_2	B	$Ae_1Be_3Ce_2A$	$\{e_4, e_5, e_6\}$
3.4	$Ae_1Be_3Ce_2A$	A	e_2	B	$Ae_1Be_3Ce_2A$	$\{e_4, e_5, e_6\}$
3.1	$Ae_1Be_3Ce_2A$	B	e_2	B	$Ae_1Be_3Ce_2A$	$\{e_4, e_5, e_6\}$
3.2	$Ae_1Be_3Ce_2A$	B	e_2	B	B	$\{e_4, e_5, e_6\}$
3.3.1–4	$Ae_1Be_3Ce_2A$	B	e_4	C	Be_4C	$\{e_5, e_6\}$
3.3.1–4	$Ae_1Be_3Ce_2A$	B	e_6	D	Be_4Ce_6D	$\{e_5\}$
3.3.1–4	$Ae_1Be_3Ce_2A$	B	e_5	B	$Be_4Ce_6De_5B$	$\{\}$
3.4	$Ae_1Be_4Ce_6De_5Be_3Ce_2A$	B	e_5	B	$Be_4Ce_6De_5B$	$\{\}$

The Euler circuit given by the modified Fleury's algorithm is $Ae_1Be_4Ce_6De_5Be_3Ce_2A$, which we more normally write as $A - B - C - D - B - C - A$ or simply as $ABCDBCA$.

Question 9.17: The Euler circuit given by the modified Fleury's algorithm is $Ae_1Be_4De_8Ee_9De_6Ce_7Ee_5Be_3Ce_2A$, which we more normally write as $A - B - D - E - D - C - E - B - C - A$ or simply as write as $ABDEDCEBCA$.

Question 9.18:

(a)
$$\begin{bmatrix} 0 & 1 & 0 & 0 & 2 \\ & 0 & 0 & 2 & 0 \\ & & 0 & 0 & 1 \\ & & & 0 & 1 \\ & & & & 0 \end{bmatrix} \text{ or } \begin{bmatrix} 0 & & & & \\ 1 & 0 & & & \\ 0 & 0 & 0 & & \\ 0 & 2 & 0 & 0 & \\ 2 & 0 & 1 & 1 & 0 \end{bmatrix}$$

(b)
$$\begin{bmatrix} 1 & 1 & 0 & 0 \\ & 0 & 0 & 3 \\ & & 1 & 1 \\ & & & 0 \end{bmatrix} \text{ or } \begin{bmatrix} 1 & & & \\ 1 & 0 & & \\ 0 & 0 & 1 & \\ 0 & 3 & 1 & 0 \end{bmatrix}$$

(c)
$$\begin{bmatrix} 1 & 1 & 0 & 0 & 0 \\ & 0 & 0 & 1 & 0 \\ & & 0 & 1 & 3 \\ & & & 0 & 1 \\ & & & & 0 \end{bmatrix} \text{ or } \begin{bmatrix} 1 & & & & \\ 1 & 0 & & & \\ 0 & 0 & 0 & & \\ 0 & 1 & 1 & 0 & \\ 0 & 0 & 3 & 1 & 0 \end{bmatrix}$$

(d)
$$\begin{bmatrix} 0 & 1 & 1 & 0 & 0 & 0 \\ & 0 & 1 & 0 & 0 & 0 \\ & & 0 & 0 & 0 & 0 \\ & & & 0 & 1 & 1 \\ & & & & 0 & 1 \\ & & & & & 0 \end{bmatrix} \text{ or } \begin{bmatrix} 0 & & & & & \\ 1 & 0 & & & & \\ 1 & 1 & 0 & & & \\ 0 & 0 & 0 & 0 & & \\ 0 & 0 & 0 & 1 & 0 & \\ 0 & 0 & 0 & 1 & 1 & 0 \end{bmatrix}$$

Question 9.19:

(a)
$$\begin{bmatrix} 0 & 1 & 1 & 0 & 0 \\ & 0 & 0 & 1 & 1 \\ & & 0 & 1 & 1 \\ & & & 0 & 1 \\ & & & & 0 \end{bmatrix} \text{ or } \begin{bmatrix} 0 & & & & \\ 1 & 0 & & & \\ 1 & 0 & 0 & & \\ 0 & 1 & 1 & 0 & \\ 0 & 1 & 1 & 1 & 0 \end{bmatrix}.$$

(b)
$$\begin{bmatrix} 0 & 1 & 1 & 0 & 0 \\ & 0 & 0 & 2 & 1 \\ & & 0 & 1 & 1 \\ & & & 0 & 1 \\ & & & & 0 \end{bmatrix} \text{ or } \begin{bmatrix} 0 & & & & \\ 1 & 0 & & & \\ 1 & 0 & 0 & & \\ 0 & 2 & 1 & 0 & \\ 0 & 1 & 1 & 1 & 0 \end{bmatrix}$$

(c)
$$\begin{bmatrix} 0 & 1 & 1 & 0 & 0 \\ & 0 & 0 & 2 & 1 \\ & & 0 & 1 & 2 \\ & & & 0 & 1 \\ & & & & 0 \end{bmatrix} \text{ or } \begin{bmatrix} 0 & & & & \\ 1 & 0 & & & \\ 1 & 0 & 0 & & \\ 0 & 2 & 1 & 0 & \\ 0 & 1 & 2 & 1 & 0 \end{bmatrix}$$

(d)
$$\begin{bmatrix} 0 & 2 & 2 & 0 & 0 \\ & 0 & 0 & 1 & 1 \\ & & 0 & 1 & 1 \\ & & & 0 & 1 \\ & & & & 0 \end{bmatrix} \text{ or } \begin{bmatrix} 0 & & & & \\ 2 & 0 & & & \\ 2 & 0 & 0 & & \\ 0 & 1 & 1 & 0 & \\ 0 & 1 & 1 & 1 & 0 \end{bmatrix}$$

Question 9.20:
$$\begin{bmatrix} 1 & 0 & 0 & 1 & 1 \\ & 0 & 1 & 1 & 0 \\ & & 0 & 1 & 1 \\ & & & 1 & 0 \\ & & & & 0 \end{bmatrix} \text{ or } \begin{bmatrix} 1 & & & & \\ 0 & 0 & & & \\ 0 & 1 & 0 & & \\ 1 & 1 & 1 & 1 & \\ 1 & 0 & 1 & 0 & 0 \end{bmatrix}$$

Question 9.21:
$$\begin{bmatrix} 0 & 1 & 1 & 0 & 0 & 0 \\ & 0 & 2 & 0 & 0 & 0 \\ & & 0 & 0 & 0 & 0 \\ & & & 1 & 1 & 2 \\ & & & & 0 & 0 \\ & & & & & 0 \end{bmatrix} \text{ or } \begin{bmatrix} 0 & & & & & \\ 1 & 0 & & & & \\ 1 & 2 & 0 & & & \\ 0 & 0 & 0 & 1 & & \\ 0 & 0 & 0 & 1 & 0 & \\ 0 & 0 & 0 & 2 & 0 & 0 \end{bmatrix}$$

Question 9.22:

$$\begin{bmatrix} 0 & 0 & 0 & 1 & 1 & 0 \\ & 0 & 1 & 2 & 0 & 0 \\ & & 0 & 1 & 0 & 0 \\ & & & 0 & 0 & 0 \\ & & & & 0 & 1 \\ & & & & & 1 \end{bmatrix} \text{ or } \begin{bmatrix} 0 & & & & & \\ 0 & 0 & & & & \\ 0 & 1 & 0 & & & \\ 1 & 2 & 1 & 0 & & \\ 1 & 0 & 0 & 0 & 0 & \\ 0 & 0 & 0 & 0 & 1 & 1 \end{bmatrix}$$

Question 9.23: The dominant operation is the Boolean addition in step 3.1.1.1.1. Thus the time complexity function is given by $f(n) = n^3$.

Question 9.24:

$$\begin{bmatrix} 1 & 1 & 1 & 0 & 0 & 0 \\ 1 & 1 & 1 & 0 & 0 & 0 \\ 1 & 1 & 1 & 0 & 0 & 0 \\ 0 & 0 & 0 & 1 & 1 & 1 \\ 0 & 0 & 0 & 1 & 1 & 1 \\ 0 & 0 & 0 & 1 & 1 & 1 \end{bmatrix}$$

Question 9.25:

$$\begin{bmatrix} 1 & 1 & 0 & 0 & 0 & 0 \\ 0 & 1 & 0 & 0 & 0 & 0 \\ 0 & 1 & 1 & 0 & 0 & 0 \\ 0 & 1 & 1 & 1 & 1 & 1 \\ 0 & 1 & 1 & 0 & 1 & 1 \\ 0 & 1 & 1 & 0 & 0 & 1 \end{bmatrix}$$

B.10   CHAPTER TEN ANSWERS

Question 10.1: $\mathcal{T}$ has 101 vertices. All 100 edges are bridges.

Question 10.2: $\mathcal{T}$ has 99 edges. All 99 edges are bridges.

Question 10.3: The new graph is disconnected and there are three components.

Question 10.4: There is one path with no repeated vertices or edges between v_1 and v_{20}. There is one path with no repeated vertices or edges between v_4 and v_{17}. There is one path with no repeated vertices or edges between v_i and v_j where $i \neq j$.

Question 10.5: The new graph has one cycle so it is no longer a tree.

Question 10.6: The vertex v_i has one parent and either one or two children.

Question 10.7: The vertex v_i has one parent and two children.

Question 10.8: The greatest number of leaves that $\mathcal{T}$ can have is 99. The least number of leaves that $\mathcal{T}$ can have is two.

Question 10.9: Any drawn graph should be isomorphic to:

Question 10.10:
(a) three; $\overline{AB}$ and $\overline{DE}$ (b) four; $\overline{AB}$ and $\overline{EF}$ (c) four; $\overline{AC}$, $\overline{BD}$, $\overline{EG}$, and $\overline{FH}$

Question 10.11:
(a) $6 \times 4 = 24$; $\overline{FG}$ (b) $4 \times 4 = 16$; $\overline{CE}$ (c) $(5 \times 4) - 1 = 19$; none

Question 10.12: Using Prim's algorithm and starting at vertex A we get the following minimal spanning tree:

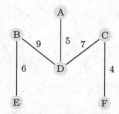

Question 10.13: Notice that in step 3.1 of Prim's algorithm we use the first edge in the edge set with minimal weight incident to exactly one vertex in $\mathcal{T}$. This means the order in which the edges are given in $\mathcal{E}$ matter. Using Prim's algorithm and starting at vertex A we get the following minimal spanning tree:

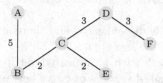

Question 10.14: The minimal spanning tree given by Prim's algorithm has vertex set $V = \{A, B, C, D, E\}$ and edge set $\mathcal{E} = \{\overline{AE}, \overline{BC}, \overline{CD}, \overline{DE}\}$.

Question 10.15: Notice that in step 3.2 of Dijkstra's algorithm we use the first edge in the edge set for which the required condition is satisfied. This means the order in which the edges are given in $\mathcal{E}$ matter. Using Dijkstra's algorithm we get the following minimal spanning tree:

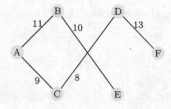

Question 10.16: Using Dijkstra's algorithm we get the following minimal spanning tree. Notice how different this spanning tree is from the spanning tree given in question 10.12.

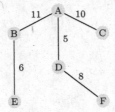

Index

Printed in the United States
By Bookmasters